AL ATHEER

IRAQI SECRET NUCLEAR SITE

ALAA M. AL TAMIMI

authorHOUSE®

AuthorHouse™ UK
1663 Liberty Drive
Bloomington, IN 47403 USA
www.authorhouse.co.uk
Phone: UK TFN: 0800 0148641 (Toll Free inside the UK)
          UK Local: (02) 0369 56322 (+44 20 3695 6322 from outside the UK)

Published by AuthorHouse 10/13/2021

ISBN: 978-1-6655-9351-9 (sc)
ISBN: 978-1-6655-9352-6 (hc)
ISBN: 978-1-6655-9353-3 (e)

Print information available on the last page.

Any people depicted in stock imagery provided by Getty Images are models,
and such images are being used for illustrative purposes only.
Certain stock imagery © Getty Images.

This book is printed on acid-free paper.

DEDICATION

To Iraq, my beloved country, the homeland of history and the future,
To my family, who brings happiness to my life.

CONTENTS

ACRONYMS

AEC: Atomic Energy Commission
AEO: Atomic Energy Organization
CIA: Central Intelligence Agency
CAFCD: Currently Accurate Complete Documentation
CNEN: the Italian Nuclear Energy Committee,
DG : Director General
CEBTP : Le Centre d'expertise du bâtiment et des travaux publics
DEA: Diplome D'Etudes Approfondies, (Diploma of Advanced Studies).
EMIS: Electro-Magnetic Isotopes Separation
HEU: Highly Enriched Uranium
IAEC: Iraqi Atomic Energy Commission
IAEO: Iraqi Atomic Energy Organization
IAEA: International Atomic Energy Agency
INC: Iraqi Nuclear Commission
INP : Iraqi Nuclear Program
INSA : Institute National Des Sciences Appliquées
INT: Iraqi Nuclear Team
GFIP: General Facility for Industrial Projects
LAMA: Active Metallurgy Testing Laboratory
NAT: Nuclear Action Team
MIC: Military Industrial Commission
MIM: Ministry of Industry and Minerals
MIMI: Ministry of Industry and Military Industrialization
NPT: Non-Proliferation Treaty
NRC: Nuclear Research Center
OMV: Ongoing Monitoring and Verification
PC-3: Petrochemical Project 3
RCC: Revolutionary Command Council
RWTS : Radioactive Waste Treatment Station:
STDAAF: Scientific and Technical Development Authority of the Armed
    Forces

SUPM : Sorbonne Université Pierre and Marie Curie.

UN: United Nations

UNMOVIC: United Nations Monitoring, Verification and Inspection Commission

UNSCOM: United Nations Special Commission

WMD: Weapon of Mass Destruction

PREFACE

Scores of books have been published in the West on the Iraqi nuclear program before Iraq's invasion in 2003. An accurate review of those publications will reveal that the main objective was to turn Western readership against the Iraqi government. Prepare them to accept war, and justify the long economic embargos imposed on Iraq that led to Iraq's deterioration of living conditions to prepare the ground for a military invasion.

The USA and UK saw this nuclear program and biological, chemical and missile programs as the declared reason for the war and the invasion of Iraq in 2003.

Against this background of justifications for that campaign, several books were published after the US-led Iraq invasion. Some were written by people who took part in the Iraq nuclear program. The purpose was to clarify the program's real nature, stages of development, and facts. These books explained the program's hidden and known aspects and nature according to their political and technical views. It was an addition needed by the Arab readership.

This book will refer to two essential themes I was directly involved. The first was designing and building the al-Atheer nuclear facility without prior experience in this field and with no foreign assistance. All other Iraqi nuclear facilities at the al-Tuwaitha, al-Tarmiya, al-Sharqat, and al-Jazira sites were designed and built by the Russians, French, Italians, Yugoslavs, and Brazilians. Such creativity was a source of weariness for many vengeful nations that despised Iraq.

The second is about passive protective plans for camouflaging the sites. These efforts were entirely successful as the forty-two-day US bombing of Iraq did not target al-Atheer. The most painful thing was the sorrowful ending. After surviving the bombardment, the site was devastated by UN inspection teams.

This book provides a brief picture of the Iraqi nuclear program and its reality, ideas, concerns, and doubts. I will not hide my personal views.

It is not objective when a witness conceals what he saw and thought. And I am not writing for pleasure and entertainment; I am writing to convey a message to the public and the generation responsible for a new quick rise in Iraq. The generation that did not experience sanctions, wars, and consequential destruction received distorted facts or did not know these facts.

The original text of a Summary of the Currently Accurate Complete Documentation (CAFCD) report on the nuclear weapons program submitted by the Iraq government to UN Security Council on December 7, 2002, is in the appendix to this book.

I hope to refresh the Iraqi memory and recultivate the Iraqi youth's patriotic spirit destroyed by the occupation and manipulated at the hands of political sectarianism.

CHAPTER 1

The Beginning

Study in France 1981–1985

It was 8:00 p.m. on Sunday, June 8, 1981. I played tennis with colleagues on a court near my residence at the College of Engineering (ENTPE) in Lyon. While we were playing, my Moroccan friend Abdul Aziz ran up to me and asked where my family lived in Baghdad. His question astonished me because I knew he knew the answer. I noticed weariness on his face. I confirmed that my family was staying in Baghdad. All my colleagues surrounded me. Abdul told me that news from Iraq was that a squadron of Israeli jet fighters had bombarded the nuclear reactor near Baghdad. The area might have been polluted by radiation; millions of the city's inhabitants might have been contaminated.

Horrified, I left the tennis court with some friends. We went to my flat to watch the news on television. All French TV channels (there were no satellite channels) stopped broadcasting regular programs and hosted specialists to discuss the environmental consequences for Baghdad and Iraq as a whole.

All were offended by the news and criticized Israel for committing such a barbaric act. The French TV channels sought to interview the Israeli ambassador and sharply criticized him. The ambassador looked helpless and did not know what to say except that Israel had to defend itself. The Israeli ambassador spoke French fluently, but that did not save him from the attacks of the journalists. All my French friends then sympathized with my citizens, who were under the severe risk of lethal radiation.

Yet something happened and made the attention paid by the French dwindle. During the main news bulletin, the correspondent of the French TV channel interviewed the Iraqi ambassador to France and asked him about the attack. The ambassador responded in Arabic, and his answer was translated into French as he did not know that language or English. My friends were shocked that my government had sent an ambassador who did not speak at least two languages. They expected him to denounce the aggression aimed at his country and a nuclear reactor provided by France, which knew its specifications and was sure it had been constructed for peaceful purposes. Most of the ambassador's responses were disappointing. As best I remember, the ambassador replied to the question, "Who said that the reactor was destroyed?" with "The men who built it are capable of rebuilding it and even better than it."

This stupid statement showed an absurd pretension of power. He may have been inspired by the political demagogy that prevailed in Iraq at that time. However, it was certain that the Iraqi ambassador saved the Israeli counterpart, who became more confident in his language. He later stated that there was no danger to Baghdad's inhabitants relying on the Iraqi ambassador's statement and denied the reactor's destruction.

After that sense of reassurance given by an Iraqi statesman, I noticed that my French friends were no longer tense. They left my flat. I watched television and listened to international Arabic broadcasting stations such as the BBC, VOA (Voice of America) and Monte Carlo until the late hours. The almost pro-Israel French media quickly seized the statement of the Iraqi ambassador; the media channels were less pointed in dealing with Israel as a result. Later on, the press lost interest in the matter and became interested in other local and international events.

The Israeli aggression on the Iraqi nuclear reactor directed my attention to the subject of the atomic bomb. I had never thought I would play a role in this dangerous industry.

The world has experienced dropping atomic bombs on cities but only twice during World War II. The United States dropped nuclear bombs on Hiroshima and Nagasaki in 1945. The two cities were devastated, causing more than 120,000 Japanese deaths in a twinkle of an eye. Others died as a result of radiation exposure. The total number of victims came to 220,000. The survivors developed cancer and especially blood cancer. Nevertheless, destructive power became a new source of energy that would serve humankind.

The decline of the global store of traditional energy sources, including oil, gas, and electricity and the environment's resulting pollution urged the search for alternatives. One was atomic energy, a historical event and a scientific revolution that took humanity to the nuclear age. Many countries started to build nuclear reactors to produce low-cost electricity.

After finishing my first year of study, I moved in September 1981 to Rennes to complete my research at the National Institute of Applied Sciences (INSA), one of the five highest institutes in France, which granted DEA Master of Science degrees.

I completed the requirements for the DEA in one academic year and moved to Paris in 1982 to join the Sorbonne doctoral program (UPMC). I did my Ph.D. research at a structural research center (CEBTP) near Paris. After four years of studying, I was awarded a Ph.D. in structural engineering.

My Return to Baghdad

I returned to Baghdad in June 1985. Iraq had been at war with Iran for five years at that time, a war forgotten by the world. It became a borderline war in which there was no victor. We remembered it when we saw placards lamenting the demise of martyrs or hearing the explosion of rockets as the Iraq-Iran war, in its fifth year, turned into a missile battle between Baghdad and Tehran.

Against this background, life for Iraqis was tragic. I noticed that Iraqis' lives had become depressing and that most Iraqis suffered because of the war, casualties, and destruction. Everyone was busy with themselves, and that was not the normal Iraqi state of mind.

Military Engineering College

I joined the instructors' staff as an assistant professor at the Department of Civil engineering, Military Engineering College, in 1985. I was assigned a task related to structural dynamics and the design of fortified structures and shelters. The latter received increasing attention during wartime.

The college of military engineering was relatively new. Established in 1973, it depended on instructors borrowed from India and East European

---

countries. The college was waiting for the return of scholars who had been sent abroad to take foreign professors' places.

In addition to my duties as an instructor, I had joined several research teams dealing with topics related to military construction and conducting studies on military facilities affiliated with the Scientific and Technical Development Authority of the Armed Forces (STDAAF). I also took part in the passive protection board's research affiliated with the Department of Scientific Affairs of the president's office.

After completing the teaching year, I moved from the Military Engineering College to the structural engineering department at the STDAAF.

Scientific Research Authority

The STDAAF, chaired at the time by Dr. Amer Mohammed Rashid, had been established in 1983.

When I started there, Amer received me in his office at the Ministry of Defense. He was a modest figure with extensive knowledge and information, which made me feel that I would be busy in my new research work.

After my time at the commission and learning about the context of the research, I was thrilled to pay back my country's benefits with the experience I had gained while earning a Ph.D.

I preceded Dr. Kwan al-Ani, a competent engineer of high character and humility, and Dr. Thamer al-Azzawi, a respected figure. I contributed to their efforts to develop the department's research action plan for 1986–1990 in the following main technical areas.

- the camouflage of structures against infrared imaging
- the effects of conventional weapons, in the particular blast and penetration effects on facilities
- design specifications for fortified structures
- the design of precast modular fortified structures
- protecting sensitive devices in fortified structures from blast waves and impacts.
- the effects of impacts on ammunition due to conventional weapons

One of the most important research topics I participated in was designing shelters to protect people from conventional and nuclear-weapon attacks. Identifying shelters' capacities was studied to achieve the security protection component for many people considering the operational aspect and providing public services appropriately and equipment maintenance.

Design standards had been established for residential buildings to consider Baghdad's conditions, especially groundwater levels.

Faw Peninsula

On the evening of Wednesday, May 14, 1986, I received a call from the officer on duty at the Armed Forces Authority of Technical Research and Development. He conveyed to me an order from LG Dr. Engineer Amir Mohammed Rasheed, chair of the authority, to attend a meeting at the authority's headquarters at the Ministry of Defense in Bab al-Muadham, Baghdad.

I met LG Amir and representatives of several departments there, including the military intelligence, engineering, air, and marine forces. Amir opened the meeting with a speech in which he informed us that our meeting came from the general command of the armed forces. We were required, he said, to submit a recommendation regarding the optimal method of destroying a bridge built by the Iranians on the Shatt al-Arab waterway after they had occupied Faw Peninsula.

After enhancing their presence in Faw, the Iranians constructed a bridge of pipes on Shatt al-Arab in Ras Albisha to supply their forces and transport heavy military equipment from Abadan to Faw.

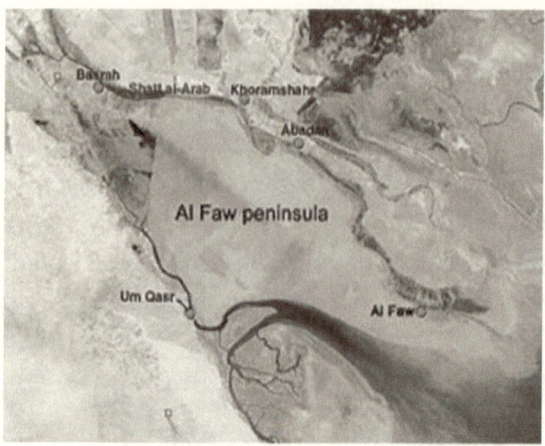

Faw Peninsula

The representative of the army intelligence showed pictures of the bridge. Amir pointed out that our mission was to find the engineering weaknesses of the bridge so we could strike and destroy it. All Iraqi air force attempts to destroy the bridge had failed though it had been accurately hit several times; the Iranians had repaired it and gotten it back into service in a short time. Amir asked me and Dr. Kawan al-Anni, head of the department, to provide our scientific perspective.

The design of the Iranian bridge was a genius approach; it was constructed under concrete and steel pipes of large diameter at the Ras al Bisha area that did not cross with the river's current. As the first row of tubes was laid, other rows were put thereupon until they reached the river's depth. Then the concrete was poured at the proper depth over the upper edge of the pipes. As such, it became a bridge prepared for military forces to cross over. The bridge could be compared to a soil dam that would not be affected by removing part of it due to its collapse in an explosion, thus the air force's failure to destroy it though it had been hit. Dr. al-Anni and I agreed that the bridge's design did not contain weak points. It was useless to continue airstrikes as it was easy for them to repair it.

At dawn on April 17, 1988, tens of thousands of Iraqi soldiers and officers carried metal mats to cross the marshes at the Mamleha (a salty region) and engage in a fight with the Iranians. These portable bridges allowed them to cross rivers and wetlands, cutting through date palm orchards.

ALAA M. AL TAMIMI

Furthermore, we had to provide wide coverings for shoes to prevent soldiers from sinking into the soggy ground during the first onslaught.

At noon on April 18, 1988, Faw was liberated, and the Iraqi flag was raised in a record time—less than thirty-six hours. It was a surprising battle in which the Iraqi military permitted the Iranian army to cross the Shatt al-Arab toward the Iranian territories by using the Iranians' pipe bridge. It was demolished and removed after the battle.

Joining the Iraqi Atomic Energy Commission (IAEC)

During this engineering and research effort, a telegram arrived on June 7, 1987, from the republic's presidency stating that I would be transferred to the Iraqi Atomic Energy Commission (IAEC).

I had to take on my new job at the Nuclear Research Center's (NRC) well-known headquarters in al-Tuwaitha near Salman Pak, southeast of Baghdad. Still, I received a phone call from Dr. Khidir Hamza, a former friend and an atomic energy officer. He told me that my site would be initially in a labor union building in Baghdad's Karadat Mariam area.

I was annoyed by my transfer and some following surprises. Still, the country's cadres' movement from one location to another was not based on personal desires but decisions made by competent authorities.

Before going into the details of my new work, I will go back a bit in history to review the development stages of atomic energy development in Iraq since 1956.

CHAPTER 2

Iraq Explores Atomic Energy

During the Monarchy

The nuclear attacks on Hiroshima and Nagasaki[1] during World War II demonstrated the destructive potential of nuclear power. Still, it also gave rise to the idea of using nuclear energy for peaceful purposes. In 1953, US President Eisenhower[2] launched the Atoms for Peace[3] program to disseminate knowledge about the peaceful uses of atomic energy. The objective was to motivate nations to avoid using the atom for military purposes.

However, the Iraqi nuclear program began with American assistance. The US administration provided Iraq with a small nuclear library and a 2 MW atomic reactor for scientific purposes during a visit Iraq's King Faisal II made to the United States in 1954.

---

[1] On August 6, 1945, during World War II (1939–45), an American B-29 bomber dropped the world's first atomic bomb on Hiroshima. The explosion wiped out 90 percent of the city and immediately killed 80,000 people; tens of thousands more later died of radiation exposure. Three days later, a second B-29 dropped another A-bomb on Nagasaki, killing an estimated 40,000 people. Japan's Emperor Hirohito announced his country's unconditional surrender in a radio address on August 15, citing the devastating power of "a new and most cruel bomb."

[2] Dwight David "Ike" Eisenhower was an American army general and statesman who served as the thirty-fourth president of the United States (1953–1961). During World War II, he was a five-star general in the army and served as supreme commander of the Allied Expeditionary Force in Europe.

[3] The United States launched an Atoms for Peace program that supplied equipment and information to schools, hospitals, and research institutions in the US and throughout the world. The first nuclear reactors in Iran, Israel, and Pakistan were built under this program.

A Republic Is Born[4]

The revolution was a turning point in Iraqi history. Its urban culture, modernization, and social order were destroyed and replaced by a military government.

The US administration changed the reactor's destination, which was sent to Iraq by sea, to Iran under the shah's rule. The Iranian nuclear reactor was constructed at the University of Tehran.

After the library arrived in Iraq, the government set up the AEC in 1956. The commission linked with the office of the prime minister. One of its purposes was to establish a scientific and technological base in cooperation with related scientific and governmental organizations such as health, education, defense, and agriculture. One of the committee's outstanding programs involved giving Iraqi staff the theoretical and practical expertise necessary to master nuclear technology.

The other development that followed was constructing a center in Baghdad by the British atomic energy research institution in 1956. The purpose was to train Iraqi staff and the other Baghdad Pact members[5]— Pakistan, Iran, and Turkey.

In 1956, Washington promised to build a research 5 MW reactor. However, the offer was withdrawn after the fall of the royal regime in Iraq in 1958 and Iraq shifting the strategic alliances to the former Soviet Union.

The IAEC decided to create the NRC in 1964 in the Shalchia district west of Baghdad. It was based in the same headquarters as the British Training Centre, which was established in 1956.

At the beginning of work in 1965, there were no laboratories necessary to conduct nuclear and radiological sciences research.

Iraq's nuclear program began in 1960 by constructing a 2 MW Soviet atomic reactor called the IRT-5000. With this reactor, Iraq started its

---

[4] In 1958, the July 14 Revolution ended the thirty-seven-year Hashemite monarchy of Iraq in a coup d'état and established the Republic of Iraq.

[5] The Baghdad Pact was a defensive organization for promoting shared political, military, and economic goals founded in 1955 by Turkey, Iraq, and Great Britain. The main purpose of the Baghdad Pact was to prevent communist incursions and foster peace in the Middle East. It was renamed the Central Treaty Organization in 1959 after Iraq pulled out of the pact.

scientific program in the field of nuclear energy for peaceful purposes. The July 14 reactor, as it was called, became operational in November 1967. The Soviet Union began to provide 10 percent–enriched uranium-235[6] and trained Iraqi experts and technicians to manage and operate the reactor and produce isotopes and production workshops.

In 1966, the July 14 reactor was about to be completed. IAEC decided to unite the reactor center, the research center, and the isotope laboratories in one organization, the NRC. In the same year, NRC moved to its permanent location at the nuclear reactor site in al-Tuwaitha

Al-Tuwaitha Nuclear Research Center

Located sixteen kilometers southeast of Baghdad, the al-Tuwaitha NRC served as the foundation of Iraq's nuclear research and development from 1966 on.

Two nuclear reactors were constructed at the al-Tuwaitha site. In 1966, a facility for radioisotope production was built with a Russian 2 MW IRT-2000 research reactor. Russian contractors upgraded the IRT-2000 to a 5 MW IRT-5000 in 1978. The site began to focus on uranium enrichment and plutonium production for a nuclear weapon in 1982.

Despite the long history of nuclear programs at al-Tuwaitha, no significant radioactive contamination from normal operations has been officially reported for the site or surrounding communities.

In 1968, Iraq signed the Non-Proliferation Treaty (NPT).[7] On October 29, 1969, Iraq ratified the treaty pledging not to manufacture nuclear weapons and agreeing to place all its nuclear materials and facilities under IAEA safeguards. The treaty gave Iraq the right to develop atomic energy but prohibited it from acquiring nuclear weapons.

---

[6] Uranium is a chemical element with the symbol U and atomic number 92. It is a silvery-grey. Uranium-235 was the first isotope that was found to be fissile.

[7] The NPT is a landmark international treaty whose objective was to prevent the spread of nuclear weapons and weapons technology, to promote cooperation in the peaceful uses of nuclear energy, and to further the goal of achieving nuclear disarmament and general and complete disarmament.

Peaceful Nuclear Programs 1966–1973

Between 1966 and 1973, remarkable achievements were accomplished. The operation and maintenance staff were completed. Laboratories were provided with equipment and material, and the NRC began research on nuclear physics, radiochemistry, radiobiology, agriculture, and isotopes. These beginnings paved the way for Iraq's entry into the nuclear science and technology arena under the flag of peaceful purposes for atomic energy.

Throughout the 1960s, Iraq didn't have any nuclear programs to direct or exploit research into or develop military nuclear programs. All academic efforts were conducted for purely scientific purposes; they supplied hospitals with isotopes instead of importing them from Britain. At that time, Iraq offered neighbouring Arab countries these isotopes as well.

In 1970, the Iraqi Nuclear Energy Commission was restructured and was headed by the minister of Higher Education and Scientific Research. The commission consisted of members of the Ministries of Health, Agriculture, Defense, Interior, Industrial, and the University of Baghdad.

The program was run by the IAEC, a small department in the Ministry of Higher Education and the Nuclear Energy Center at the Atomic Energy facility at al-Tuwaitha.

The October Arab-Israeli War[8] of 1973, known to Israelis as the Yom Kippur War and to Arabs as the October Liberation War, impacted Iraq's leadership. The media played a role in spreading this concept after the war was stopped though the Arab armies had progressed. It was said that Israel threatened to use nuclear weapons if the Arab armies did not stop advancing. Besides, Iraq had problems with the shah of Iran, which supported the Kurdish rebellion in northern Iraq. The shah sought to possess advanced nuclear reactors.

---

[8] The 1973 Arab-Israeli War was fought from October 6 to October 25, 1973, by Egypt and Syria against Israel. The war took place mostly in Sinai and the Golan—occupied by Israel during the 1967 Six-Day War—with some fighting in Egypt and northern Israel. Egypt's initial war objective was to use its military to seize a foothold on the east bank of the Suez Canal and use this to negotiate the return of the rest of Sinai

These developments pushed the Iraqi leadership to think of a strategic deterrent force as potential enemies threatened it.

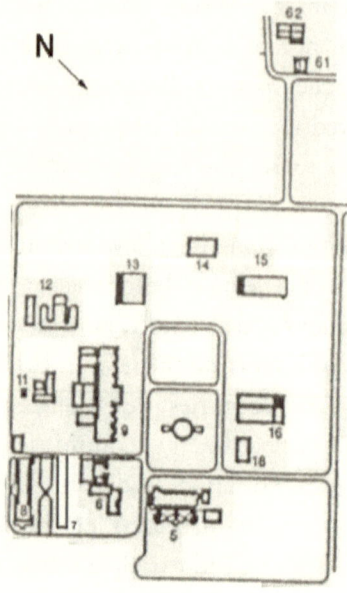

al-Tuwaitha Nuclear Research Center 1970[9]

Iraq Atomic Energy Commission (IAEC) 1974–1982

In light of the abovementioned facts, a new law was adopted in 1974. The IAEC was set up to be headed by the Revolutionary Command Council (RCC)[10] under Saddam Hussein.

A strategy was formulated to develop Iraq's nuclear potential to match its capacities and regional position. As such, the IAEC prepared a program to possess nuclear fuel cycle technology.

The nuclear fuel cycle started in 1976 by discovering radioactive

---

[9] The development of al-Tuwaitha: "What If the Public or the IAEA had Overhead Imagery?" by David Albright, Corey Gay, and Khidhir Hamza, April 26, 1999, Institute for Science and International security.

[10] The Iraqi Revolutionary Command Council was established after the military coup in 1968 and was the ultimate decision-making body in Iraq. It was officially dissolved on May 23, 2003 by Paul Bremer Order, administrator of the Coalition Provisional Authority following the 2003 invasion of Iraq by the United States.

material in Iraq and extracting uranium from crude phosphate, which was abundant in Iraq's western desert at the Akashat mine, 420 kilometers west of Baghdad.

Simultaneously, a contract was signed in 1976 with Sebetra, a Belgian company, to exploit phosphate extracted from the Akashat mine. (Phosphate is used to produce fertilizers.) In 1979, a Swiss company built a uranium ore production facility in the region at al-Qaim City .

The nuclear program expanded in the 1970s by acquiring Osirak, an Osiris-class research reactor,[11] from the French government. On November 18, 1976, this deal was finalized when Iraq signed a contract with French companies with the French Atomic Energy Commission (CEA). Construction on the new reactor began in 1979.

The aim was to construct the July 17 project, which consisted of a Tammuz-1[12] (July 1) reactor with a 40 MW capacity. It was similar to the French Osiris reactor near Paris.

The reactor could be used to produce isotopes. It is to be noted that the fuel used to operate this reactor was a kind of a meld of uranium and aluminum-containing 93 percent enriched uranium. There was also a small reactor called Tammuz-2 (July 2) Isis with a 500-KW capacity.

In 1979, Iraq contracted with the Italian company SNIA-Techint for a pilot plutonium separation and handling facility and a uranium refining and fuel-manufacturing plant.

Another agreement was signed on January 30, 1976, between AEC and CNEN, the Italian Nuclear Energy Committee. The agreement stipulated that an Italian company would build a project called then Tammuz-30 (July 30). It consisted of a laboratory to manufacture fuel for energy-producing reactors and included a technology hall designed for chemical engineering research, testing labs, and small mechanical and electrical workshops. The project contained a laboratory for producing radioactive isotopes, medical diagnosis instruments, and material testing labs.

A small plutonium-separation program started in the mid-1970s.

---

[11] Oiris, from the Greek for Us-yri, he who sits on the throne, i.e. the king, was a French nuclear research reactor (70 MW) operating since 1966.

[12] Tammuz is the sevenths of the twelve months of the Babylon calendar; its name came from an ancient Mesopotamian god associated with shepherds who was also the primary consort of the goddess Inanna, later known as Ishtar.

Following contact with SNIA-Techint of Italy, a facility was established in Baghdad to research fuel processing under IAEA safeguards. This laboratory was eventually able to separate small quantities of plutonium.

Consequently, there was a plan to construct an electron-nuclear station to produce 400 MW. For that purpose, the IAEC entered negotiations with Germany, France, Britain, Italy, Japan, Canada, Belgium, and Finland. There was a tendency to enter a contract with Finland. Negotiations went on till the mid-1980s. In parallel, studies and tests were conducted to choose a good site for the station.

Iraq's nuclear programs continued in public and were known by the IAEA. Iraq did not have any plan to possess a nuclear reactor for military purposes. Indeed, most Iraqi AEC projects up to 1981 were built by global companies.

al-Tuwaitha (the headquarters of the AEC) as the July 14 reactor and service facility was constructed by the Soviet Union in 1964. The July 17 reactor and its labs had been built by French companies (1976–1981), and the Italians built the July 30 project in 1976.

In April 1975, the talented nuclear physicist Dr. Jafar Dhia[13] Jafar joined the AE and quickly took a leading role in the atomic weapons program.

In addition to being a brilliant scientist, Jafar hailed from Iraqi society's upper crust, his father having been a cabinet minister under the last monarch.

Raids on the al-Tuwaitha Nuclear Site

The first air raid to prevent nuclear weapons development happened eight days after the Iran-Iraq War[14] began in 1980. Aircraft carried out the first airstrike on a nuclear reactor and the first preemptive airstrike to prevent any country from developing nuclear weapons capabilities.

---

[13] Considered the father of Iraq's nuclear program.

[14] Started September 22, 1980 and lasted till August 8, 1988; it resulted in at least half a million casualties and several billion dollars' worth of damages but no real gains by each side.

Iran Scorch Sword Operation

At dawn on September 30, 1980, four US-made F-4E Phantoms flew low over central Iraq, each loaded with air-to-air missiles and three thousand pounds of bombs.

Moments later, two Phantoms peeled off and dropped to a lower altitude to avoid radar detection after crossing the Iraq border. They were flying a stealth mission to al-Tuwaitha, sixteen kilometers southeast of Baghdad, to strike an incomplete nuclear reactor. Only secondary buildings were hit. The attack damaged pipes, cooling pumps, and lab facilities. Hundreds of French and Italian technicians withdrew from the building after the raid though some eventually returned.

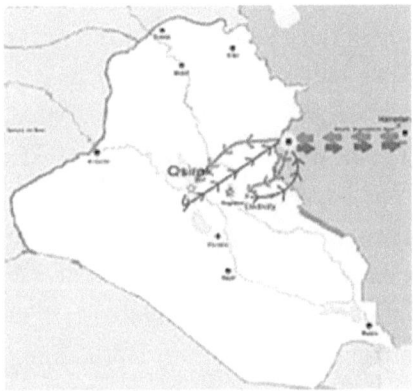

Iran's Scorch Sword Operation

Israel and Peaceful Iraq's Nuclear Program

Israel applied a multipronged strategy to halt the reactor's construction—diplomacy, a media campaign, sabotage, assassination, and military action.

Israel launched an open war to prevent the delivery of French reactors to Iraq, trace the personnel, and follow up on Iraqi nuclear program workers.

As diplomacy stalled, Israel turned to sabotage and assassination. Israel's first attempt to disrupt Iraq's plans occurred on April 6, 1979. Two reactor cores lay in storage at a plant at the French firm of Constructions Navales et Industrielles de la Méditerranée in La Seyne Sur-Mer near Toulon awaiting shipment to Iraq.

A Mossad operation known as Operation Sphinx smuggled in seven operatives. They placed five explosive charges on the cores. They detonated them, damaging the cores and setting back Iraq's program by at least half a year.[15] The most severe damage to the cores was repaired. Still, X-rays revealed hairline fractures throughout the core of Osirak that could not be fixed without completely rebuilding it, a process that would take two years.

Significant parts of the reactor were destroyed but repaired. The shipment was delayed for another six months. The two reactors with the nuclear fuel required for operating the Tammuz reactor were shipped to al-Tuwaitha.

Threats followed by bombings directed at Italian and French companies providing laboratory equipment for the nuclear program also failed to deter exports.

After the sabotage of the reactor core, Israel's next target was a person, not an object. Mossad kept a team operating in France to continue its assault on the Iraqi project after the bombing at Toulon. Yehia al-Meshad[16] was a respected Egyptian nuclear engineer hired by Iraq to supervise the reactor deal. al-Meshad had a role in the program but was not a key figure in it. On June 13, 1980, al-Meshad was stabbed to death in his room.

Throughout the rest of 1980, a deadly harassment campaign continued against the Iraqi program. On August 7, 1980, the office of SNIA-Techint, the Italian firm that supplied plutonium reprocessing technology to the Iraqis, was bombed.[17] Two other Iraqi nuclear engineers perished in suspicious circumstances.

An electrical engineer named Salman Rashid,[18] who was working on separating the electro-magnetic isotopes (EMIS) project, suddenly became very ill while on a two-month trip to Geneva to work with Brown Boveri. He died of a mysterious ailment ten days later. Six months later,

---

[15] Raviv and Melman 1990, 251; Ostrovsky and Hoy 1990, 19–20.

[16] Yahya al-Mashad (1932–1980) was an Egyptian nuclear scientist who was killed in a Paris hotel room in June 1980.

[17] Gsponer and Hurni 1995, 21.

[18] According to Swiss physicist André Gsponer, who worked for CERN, a European institute for the study of particles near Geneva, Iraqi engineer Dr. Salman Rashid al-Lami had come to the institute in 1979 to study electro-magnets supposedly for energy storing purposes.

on December 13, 1980, another engineer, Abdul-Rahman Abdul Rasool, died of poisoning at a French banquet.[19]

Israel resorted to another more dangerous option—a military solution. On June 7, 1981, Israeli aircraft dropped several bombs on Osirak and scored enough hits to knock out the reactor permanently.

Israel's Operation Opera

The Israeli Opera operation involved F-16 and F-15 fighter jets launched from a military base in the Israeli-controlled Sinai Peninsula. They crossed the Saudi border at a very low altitude (under 150 meters), flew 1,300 kilometers before reaching the target, and unloaded their cargo at 5:30 p.m. on Sunday, July 7, 1981. The operation lasted approximately two minutes; sixteen missiles were fired, fourteen of which hit their targets. Two bombs missed the reactor but hit a thirty-meter tunnel connecting the reactor to a lab.

The raid destroyed the Tammuz-1 reactor but did not collapse the reactor building though it severely damaged it. The small auxiliary reactor, which was full of nuclear fuel, was severely damaged but fortunately did not leak. If a leak had occurred, the consequences of radioactive contamination would have been catastrophic.

Israel's attack was carried out at sunset. There were several advantages to that timing. It minimized the opportunity for Iraqi air defenses at the site to detect them visually. The target, however, was easy to spot. The near-horizontal sunlight illuminated the light-colored dome for the approaching F-16s. In addition, if any aircraft were lost, the Israelis would have been able to conduct search-and-rescue missions under cover of darkness. All the planes returned to their base.

The sudden attack shocked the leadership and people of Iraq, the Arab world, and Third-World countries. It was a lethal strike on Iraq's ambition to possess nuclear technology. The operation had been planned since 1976 when Iraq and France had signed an agreement to build the two reactors.

---

[19] Hamza and Stein 2000, 133–34.

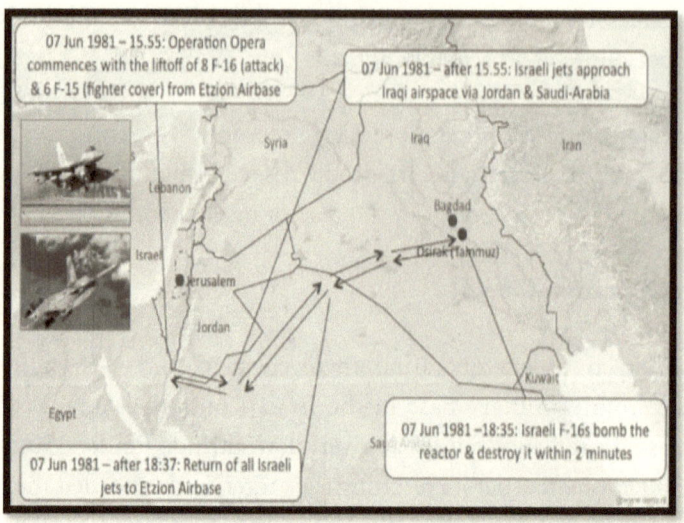

Israel's Operation Opera

The workshop buildings, LAMA and RWTS radioactive waste treatment plant, were hit. The result was severe for the reactor building and buildings nearby. Eleven people, including a French technician, were killed in the bombing. This Israeli aggression terminated this phase of developing a peaceful nuclear capacity. With the support of influential nations across the globe, the UN criticized the attack as a "deliberate and unprecedented act of aggression."

In response to the attack, the UN Security Council adopted Resolution 487, which was unanimously approved, criticizing the attack; the council considered it a violation of the UN charter and international standards of conduct.

Despite Iraqi and French claims to the contrary, Israel believed that the facility had been designed to produce nuclear weapons and called the bombing an act of self-defense.

The United States supported the resolution because Iraq was an ally against Iran. Within seconds, Israel terminated an international scientific program on Iraqi territories. It devastated the NRC, which was working within the framework and guaranties of the IAEA.[20]

After the destruction of Osirak, Iraq initially attempted to replace it.

---

[20] The International Atomic Energy Agency.

Saudi Arabia offered to finance a replacement. French President Mitterrand declared an in-principle agreement to rebuild Osirak after consultation with Iraqi deputy prime minister Tariq Aziz in August 1981. However, France wanted to tighten its controls on the project, including on a reactor core surveillance system.[21]

Uranium Enrichment Efforts

After months of Israeli aggression, in June 1981, Iraq started the secret nuclear program when it turned out that international conventions were valueless despite the transparent procedures it had adopted. This unjustifiable act of aggression prompted Iraqi authorities to pay attention to a deplorable international position in which a state could throw aside international laws without penalty.

The Iraqis realized that the project was not protectable and would be unsuccessful unless it adopted measures to keep it hidden and not rely on foreign help or financing.

In the wake of the Israeli strike, the Iraqi ambition to build a nuclear research center revealed a weakness—the site had been successfully attacked. The Iraqis realized that it would be difficult to conceal a nuclear reactor in a world of satellites and aerial surveillance. They were alerted to the importance of minimizing their facilities' risk of being targeted for destruction through passive protection measures.

Iraqi scientists became more cautious and started to develop concepts and procedures to protect the staff of the nuclear program after the series of assassinations.

Confusion in the Program

Throughout the 1980s, the IAEC continued to acquire nuclear technology with international companies at al-Tuwaitha, the site declared to the IAEA. It worked on undeclared nuclear sites through the cover of state construction companies contracting with global companies.

Regionally, Iraq was between the jaws of Iran's shah's pliers to the east and Israel to the west.

---

[21] Evron 1994, 28.

Iran established the Atomic Energy Commission in 1973, and its chair was deputy prime minister. Shah Mohammad Reza Pahlavi had an ambitious plan to raise Iran's nuclear reactor capacity of 20,000 MW by 2000 and to build two reactors (Bushehr 1 and 2) with a total of 1,200 MW supervised by German company KWU and two other 900 MW reactors created by the French company Farm Atom.

Under the circumstances prevailing in Iraq at that time on regional and global levels, there had been a tendency to build an Iraqi defense force. Nuclear weapons were viewed as an essential element of that force. The decision to create nuclear armaments, known as the Iraq National Program, began in fall 1981.

Jafar narrated his meeting with President Saddam Hussain in 1981.[22] In that meeting, Saddam directly instructed him, saying, "We must build a nuclear program with Iraqi hands and minds. We have to possess nuclear weapons and own the foundations of nuclear technology."

"I will do my best, sir, Mr. President," replied Dr. Jafar.

The program began. Senior engineers and others were transferred to work for the program despite their desires because the atomic program was paramount.

The war with Iran, which lasted from 1982 to 1987, took a considerable toll on the economy, but strategic sectors were shielded to support the war effort.

Technical Options

The theory of uranium enrichment was well known; it had been published and taught. Still, putting the idea into practice offered challenges depending on a country's circumstances, capabilities, and constraints.

Iraq faced more constraints than most countries when it developed a nuclear weapons capability; these included a severe shortage of skilled human resources and an underdeveloped industrial capacity. In addition, it could not elicit foreign assistance that could alert other countries to the program's true nature and provoke Iranian or Israeli aggression.

In the second half of 1981, there was no attempt to explore

---

[22] In chapter 2 of his book The Last Confession—The truth of the Iraq nuclear program.

weaponization or develop a weapon design; efforts focused on producing enriched uranium.

There are two general options for producing fissile materials for military or civilian purposes—plutonium breeding or enrichment. Plutonium breeding is the cheapest and fastest option for a small or a developing nation with limited military nuclear ambitions without any technical or political obstacles. It requires large-scale buildings for the reactor and its attachment in one site.

The enrichment option was chosen due to the absence of a suitable reactor in the country and the threat of future attacks since the plutonium route was easy to detect by hostile monitoring devices. It would be an open target for aggression compared to the enrichment option, which allowed the building laboratories, workshops, and scattered locations.

The question that remained was which enrichment technology should be implemented in Iraq. The initial assessment focused on four options: laser isotope separation, gaseous centrifuge, gaseous diffusion, and electro-magnetic isotopic separation (EMIS). The primary options were the EMIS method and gas diffusion developed in the US during World War II. Laser isotope separation was not a proven method for producing sufficient quantities of enriched uranium. The country lacked the basic know-how and the technological capacity to pursue this method. Gaseous centrifuges were deemed too difficult. This method required unique materials that could not be produced locally. The engineering challenges associated with the centrifuges' high speed were considered too demanding given limited know-how, and that centrifuges could not be employed without foreign assistance.

Gaseous diffusion, where uranium hexafluoride was separated through a porous membrane, was an old and proven method, but this approach still presented considerable challenges. First, the NRC lacked the basic technical know-how and industrial capacity to produce the barrier material. The machines required were under export controls and could not be built locally. The IAEC also judged this to be an uneconomical method requiring thousands of cascades of porous membranes to separate the enriched uranium from other compounds.

Between 1982 and 1985, there was no direct pressure to produce weapons. No one had started to think about what kind of weapons they ought to be and how they would be delivered. The roadmap was based on Jafar's memorandum in 1981, which focused on technological options and assessments rather than deadlines or benchmarks.

The Clandestine Program

In 1982, the IAEC staff was around 200. Over the next three years, the number of departments and staff increased to cope with specific tasks. The NRC continued research activities and international publishing to reduce foreign suspicion, but those activities were completely separate from the clandestine program. The NRC continued to conduct primary research consistent with previous chemistry, physics, biology, agriculture, medical diagnostics, and radioisotopes programs. Their findings were communicated in scholarly publications and at academic conferences.

Iraq Nuclear Program Expands

In 1982, the nuclear program linked up with the personal secretary of the republic's president. Figures and codes substituted for the actual names of departments and offices. The chair of the IAEC was Izzat al-Dorri, who served in his capacity as the RCC's vice-chair. The vice-chair was Dr. Humam. The organization included six departments, Head Office (1000), Office of Science Policy (2000), Office of Studies and Development (3000), Office of Administration and Engineering Services (4000), Office of Projects (5000), and the Nuclear Research Center (6000).

Jafar and Humam were responsible for the program. The IAEC commissioners oversaw the management and the project. The program resembled other state agencies; it underwent frequent reorganization, rapid institutional growth, and intra-agency competition.

Office 3000 oversaw several crucial scientific and technical development areas, including exploring options for uranium enrichment, finding relevant research development, selecting locations for labs and industrial buildings, and setting up procurement units in other state organizations to avoid foreign detection.

Exploring the Fuel Cycles

During the 1970s, geologists and chemists revived earlier efforts to search for and exploit uranium deposits in northern Iraq. Interest in producing yellowcake[23] from natural uranium (i.e., refining uranium) became a priority.

In 1979, the Ministry of Industry and Minerals (MIM) established a fertilizer production complex at al-Qaim. In May 1982, MIM (on behalf of the AEO) commissioned a Belgian company to build a unit to extract yellowcake[24] from phosphoric acid at Ukashat. In January 1985, this unit became operational. The next step in the process was to convert the yellowcake to a suitable feed material for EMIS. During 1980–1981, the AEO finalized contracts to move the program from initial research and development toward production.

In the 1980s, AEO signed a contract with NATRON, a Brazilian company, to build a uranium purification facility. When Israel bombed the July 17 reactor, the AEO decided to depend on the Brazilian company's design. This design was modified, and additional equipment was purchased to suit it; the design modifications were completed in summer 1987. A site north of Mosul for the conversion plant, named al Jazira, was selected in 1989. The plant processed uranium dioxide (UO_2) and uranium tetrachloride (UCl_4) from mines in Akashat.

Electro-Magnetic Isotopes Separation (EMIS)

This method was described in reports from the project gifted to the NRC library in the 1950s. The technique had been around for many years; the US used it between June 1944 and July 1945 to make the nuclear charge for the bomb dropped on Hiroshima. In principle, the technique involves separating with the help of a powerful magnet atom of uranium-235 (military grade) from the atoms of uranium-238 (poorly fissile) that make up natural uranium. The machine, developed at the

---

[23] A uranium concentrate powder that can be further refined into nuclear fuel or undergo enrichment.

[24] A uranium concentrate powder that can be further refined into nuclear fuel or undergo enrichment to become uranium oxide and pure uranium tetrachloride.

University of California, was known as a Calutron, a contraction of California University Cyclotron.

The details of the technique were made public between 1946 and 1956. Of the several possible methods for obtaining enriched uranium, the INC decided to use the calutrons technique after modernizing and perfecting it.

Exploring Enrichment

In the early 1980s, the EMIS was a peaceful research program that provided fuel for its Iraqi power reactor and research reactor projects. After the Israeli attack in June 1981, the EMIS program's goal changed. It built two production units to achieve fifteen kilograms per year of weapons-grade uranium using natural uranium feed.

In 1987, after several years of research and development work with mixed success, the IAEC adopted a plan to distribute and construct production systems in new locations away from al-Tuwaitha to lessen the potential destruction attacks could constitute. Sites began preparations for the production stage in tandem with the first and second series of projects.

The distribution of sites conflicted with the need for scientists and engineers to be physically proximate during the research and development process, mainly as these ran in tandem. They also spread out workshops that added to the cost, which caused delays and required duplicating human resources and machines. Sites involved in uranium enrichment technologies were spread around Baghdad. Duplicate sites and facilities were to focus on the extraction and refinement of uranium feedstocks in northwestern Iraq.

The Iraqi nuclear weapons effort received raw uranium from mines at Ukashat. Seven facilities were prominent in the calutron enrichment program.

The electro-mechanical and electronic workshops in al-Tiwathah were transformed from small workshops into integrated factories. An industrial area was built in Zafaraniyah, south of Baghdad, to support the nuclear program and develop high-speed centrifuges to enrich uranium by electro-magnetic isotope separation (EMIS).

The uranium enrichment production plant was divided into two identical facilities. The al-Tarmiya project, 35 kilometers north of Baghdad, was built by a Yugoslav company. The al-Sharqat project, 220 kilometers northwest of Baghdad (a replica of the Tarmiya site), was built by the Fao General Establishment. Each produced fifteen kilograms of highly enriched uranium yearly.

The al-Sharqat project was viewed as a second production site that would simultaneously occur as the al-Tarmiya project.

Reorganization and New Groups

In summer 1987, Research and Development (3000) was restructured by dividing into three groups.

Group 1 (G-1) included all enrichment activities according to gaseous diffusion technology, barrier development, separation process, and control and instrumentation design. It was transferred by the end of 1987 and attached to Hussein Kamil, the minister of Military Industrialization. Its name changed to the Center of Engineering Designs.

Group 2 (G-2) included all enrichment activities according to the electro-magnetic separation technology, design, and operation of chemical projects, engineering design, and material science. It handled the local production of natural uranium compounds and industrial-scale facilities to produce pure uranium compounds suitable for fuel fabrication or isotopic enrichment.

Research and development of the full range of enrichment technologies culminated in the industrial-scale exploitation of EMIS.

Group 3 (G-3) included all support activities, including planning, purchasing, administration, general engineering support, mechanical fabrication, equipment, utilities, information, and technical support.

The electro-mechanical and electronic workshops at al-Tuwaitha were transformed from small workshops into integrated factories. An industrial area was built in Zafarania, south of Baghdad. The industrial complex contained gigantic production workshops, stores, and administrative premises. The total area of the site was about 60,000 square meters.

The Zafarania complex consisted of two big factories. The Dijlah factory handled electrical and electronic products. It was an advanced technology center with high-quality output. The al-Rabie mechanical products factories rose to produce automatic operation machines. The factory had a giant plasma cutting machine of 12 meters x 3 meters up to a thickness of 70 millimeters, one of Iraq's biggest cutting machines.

CHAPTER 3

Steps Toward Weaponization

In 1987, the IAEC was replaced by the Iraqi Atomic Energy Organization (IAEO). Dr. Humam Abdul Khaliq was appointed as chair, and Dr. Jafar Dhia Jafar was its vice-chair. The link between the organization and the RCC vice-chair, Ezzat al Duri, was terminated.

The IAEO chair submitted a study to the presidency office edited mainly by Dr. Khidir Abdul Abbas Hamza.[25] The study dealt with the weaponization of nuclear weapons. This development could be made possible by constructing a new location out of al-Tuwaitha. The site should be fortified and subject to confidentiality. The project required the support of many institutions affiliated with the Authority of Military Industrialization (AMI).

The president referred the study to Hussain Kamil, chair of the Authority of Special Security (ASS), to elicit his opinion. Kamil suggested that this program should be separated from AEO and linked with the Special Security Authority. The Military Industrialization Authority (MIA) was entrusted with implementing the program because part of the weaponization program was related to the explosives industry. The purpose was to give the IAEO the power to dedicate its staff to carry out enrichment programs.

Kamil recommended making the weaponization project his responsibility, and the president agreed. Kamil's new portfolio included all arms production, including WMDs (chemical, biological, and nuclear weapons programs). This weakened the IAEO's influence over the sensitive dimension of the program.

---

[25] Dr. Khidir Abdul Abbas Hamza left Iraq in 1995 and was given asylum by the United States.

As the program moved closer to the threshold of a nuclear weapons capability, the entire reign of the IAEO was about to end.

Kamil transferred a dozen scientists and engineers from al-Tuwaitha to ASS. Dr. Khidir Abdul Abbas Hamza, the head of the group, was transferred along with Drs. Ghazi al-Shahir, Salah al-Khafaji and several physicists and engineers. Next to Hussain Kamil's office, the trade unions building was ordered to be the new project's personnel's headquarters.

I joined Dr. Hamza's group on June 19, 1987, and participated with tens of Iraqi scientists, engineers, and experts of different specializations. The Iraqi nuclear program was a scientific and an industrial renaissance project, not a project to produce just deterrence.

On my new job, I asked Hamza to point out the reason for transferring my services to this group and what would be required. He replied that my work involved the same engineering tasks—analysis, structural design, and fortified structures and shelters. He suggested that I review the Manhattan Project, whose references were available at IAEO library, which I did. Library held scientific publications on the Manhattan Project issued at the end of the 1940s and presented to Iraq as a gift during the monarchy.

I had reviewed reports and publications of the US technical information known by the name of TID, which was concerned with the US project to manufacture the atomic bomb. The Manhattan Project's confidentiality was removed during the 1970s. I concluded that required laboratories were involved with nuclear weaponization when enrichment projects were completed. My immediate mission was to design a laboratory that would endure repeated external explosions of cycloramic charges of 15 kilograms at a distance of fifteen meters from the laboratory's front wall. The lab was to withstand a 200-kilogram explosive at a distance of thirty meters from the lab's external border. The charges to be used consisted of highly explosive PETN. There was an opening with a forty-centimetre diameter in the front wall through which cameras would film the explosions.

I should design and construct another lab that would endure the effect of recurrent "contained explosions" of a one-kilogram TNT charge. I had not created anything that would tolerate such internal and external explosions; no publications dealt with those requirements. Even if they were available, they would have been confidential. It was impossible to contact international consultants to prepare the designs; such a request would reveal the actual purpose behind the establishment's creation. My experience with shelter design and designing labs for testing explosives differed.

Shelters were required to endure direct and indirect strikes. The lab had to remain accommodating and could not be affected by repeated explosions. It was not an easy task. From an engineering perspective, the lab had to be designed to be flexible enough to withstand great explosions to protect it and its personnel. I was annoyed by this shift and a series of ensuing surprises, but it was out of my hands. I thought of pulling out because I did not feel competent enough to design such a lab considering my lack of accurate information with which to do so. I conducted as much research as I could; I read all the scientific and engineering resources I could find. I was convinced that I had no alternative but to withdraw to expose myself and my family to the regime's wrath.

My team and I began to study the literature to define the challenges we faced, assess what resources would be required, and consider what help we could mobilize. We had not received any specifications for basic design, and we lacked relevant expertise.

I decided to go ahead with the design and bear the consequences. I requested several engineers to assist me. Architect Esam and structural engineer Yunis, two highly qualified engineers, joined my team. They had played a significant role in our achievements and had shared challenging missions with me. Later on, other engineering team members joined us, including Jabbar, Dhia, Muaed, and Sabah Hadi.

Then I had to select a site location. After repeated visits to specific sites accompanied by my technical team, I proposed a place close to Hiteen Enterprise, affiliated with MIA in the district of Jurf al Sakhar. The enterprise had a firing range and dealt with missile development. For camouflage, I was careful to choose the labs' locations in a way that would mimic Hiteen Enterprise's designs. I intended to separate our project from the Hiteen establishment by a wire fence, not a brick wall. The purpose was to convince any air surveillance that these labs were part of the Hiteen establishment

The proposed site was approved, and we called our program Project 100.

Project 100

Project 100 included a high-explosives test bunker used for hydro-dynamic experiments and three warehouses with all other internal roads, administrative buildings, external fences, water, and electrical services.

My design team started to work immediately on collecting the design parameters, including the required lab area and the equipment to be installed in each laboratory. The purpose was to absorb the vibrations explosions would cause. Dr. Khidir could not tell us all the required elements as the equipment had not yet been bought. He informed me that he would try responding to my other technical inquiries after returning from abroad. He was travelling with Drs. Saleh al-Khafaji and Ghazi al-Shahari to purchase some equipment for the project.

They visited Germany during the summer of 1987 to obtain equipment from various companies. They contacted Leybold-Heraus, Balzers, Degusa, Leyda, Klockner and Laser for the following items:

- Mirror machinery techniques and machinery.
- Melting furnaces for U.
- Purification melting zone furnace for U.
- Vacuum systems.
- Vacuum furnaces.
- Microscopes.
- Mass spectrometers.
- Mechanical properties testing facilities.
- Fast oscilloscopes.
- Special additives for concrete and coating material.
- Spherical machining facilities.

Three weeks later, I met Dr. Khidir. He looked worried and absentminded. After a while, he did not show up at his office. Then he was removed from his post; he returned to his former job at AECO.

The design represented a challenge for us. Designing and constructing a building that could stand up against external explosions without approved engineering calculations could lead to catastrophe. I asked my design team to visit the explosion labs at some establishments affiliated with MIA to benefit from an idea in designing that East European companies had implemented. That, however, did not give us much information.

Bunker Building

My engineering team and I decided to prepare the architectural and construction design for the lab. We entered complicated and detailed debates with the physicists' team to establish the lab's required area. We had to decide on the location and height of camera openings and the distance and the sizes of the charges we would explode outside the lab, but we still did not know the equipment's characteristics.

The critical factors included the placement of the explosives and the time lag between explosions for each experiment using cameras that could shoot pictures in nanoseconds. Two cameras could be installed in various positions. One was to film the blasts frame by frame. The other camera, called a streak camera, was to take successive shots with open lenses to record what happened during the explosions. The process required installing a third camera, a flash X-ray camera linked with a high-frequency analyzer, to receive signals from the uranium ball's sensors. The purpose was to determine the explosive pressure in every experiment.

Due to its transient nature, the blast load led to severe damage in structures different from other loadings such as earthquakes and wind. Peak pressures were much higher than the static collapse load of the design. Still, their durations were generally short compared to natural periods of structure components. The approach to designing a structure capable of surviving the effects of high-intensity but short-duration loads had to be different from those adopted for conventional designs. Designers had to try to reduce these effects by unique methods. This matter pushed us to think of using unconventional reinforcements distribution called lacing reinforcement, which would increase the ductility levels of structural elements. This was a new method employed for the first time in Iraq. No one had heard about it in Iraq before.

Conventional Reinforced Cement Concrete RCC is known to have limited ductility and confinement capabilities. These properties are primarily required for structures present in the blast-loading environment.

The structural properties of RCC can be improved by modifying the concrete matrix and by suitably detailing the reinforcements.

The structure had to be designed as a flexible system allowing its joints to deform considerably. The stresses in the elements must go beyond the elastic limit so that the available strength in the post-yield region is fully utilized. An elastoplastic model for concrete design was adopted. The design approach also used the enhanced yield strength due to the high strain rates possible in blast and impact loads. The low probability of occurrence and repetition of shock loads allows the designer to utilize the energy-absorbing characteristics of a carefully designed structure. This meant that the structure should be engineered to be as ductile as possible.

List of the primary references used to Design external and Internal explosive laboratories

1. Annals of the New York Academy of Science (1986)
2. Baker, W. E, Explosion in air (1973)
3. A manual for the prediction of blast and fragment loading on structures, Baker, W. E. U.S. department of energy, Amarillo, Texas, (1980)
4. Bodurtha, F.T., Industrial explosion prevention and protection, New York, McGraw-Hill book company (1980)
5. Design of structures to resist the effects of atomic weapons TN 5865, U.S. (1965)
6. The dynamic of explosion and its use, Amsterdam Elsevier scientific publications (1979)
7. Structures to resist the effect of accidental explosion (1996), Department of army technical manual TMS-1300, U.S. (1969).
8. Suppressive shields structural design and analysis handbook, U.S. army corps of engineers (1977).

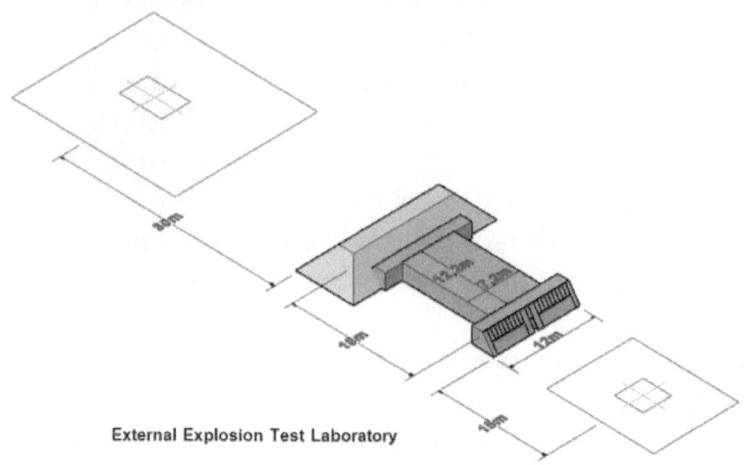

External Explosion Test Laboratory

External explosion laboratory

Administrative Changes

In Dr. khidir's book *Saddam's Bombmaker*,[26] he wrote that Hussain Kamil[27] asked him to complete the project within two years—by 1989. However, he told Kamil that the project required enriched uranium. The matter lay within the competence of the AECO, which needed five years to provide it. According to Khider, Kamil was furious. He asked him to submit a report to the president showing the failure of the AECO to keep its promise to prepare depleted uranium on time.

Nobody knew what happened. There were many narratives. However, Dr. al-Nueimi, in his book,[28] stated,

Dr. Khidir submitted a report to Lieutenant General Hussein Kamil. He specified that a weaponization program would take a decade and require some three hundred specialists and a billion-dollar budget. He pointed out the necessity of meeting specific requirements, including the staff, finance, and the period necessary to achieve the project.

---

[26] *Saddam's Bombmaker* by Khidir Hamza (with Jeff Stein).
[27] After his 1995 defection to Jordan, Hussain Kamel characterized him as a professional liar; www.informationclearinghouse.info/pdf/unscom950822.pdf.
[28] *The Last Confession, the Reality of the Nuclear Program* (2005), coauthored by Dr. Dhia Jafar.

The estimate for the financial investment required for the infrastructure was very high, totaling more than one billion USD. the initial time estimated to complete the program was about ten years. The total number of qualified personnel required was also very high. Consequently, a request was made by PMIC to have these estimates assessed by IAEC. As a result of this assessment, the Dr. Khidir team's activities were terminated. The personnel returned to the IAEC in Nov. 1987.

In his turn, Hussein Kamil submitted the report to President Saddam. In September 1987, Saddam's secretary Yousef Hammadi called specialists at AEO and Special Security Department to discuss Hamza's findings. It was attended by Dr. Humam, Dr. Jafar and Dr. Khalid Saeed, the Nuclear Research Centre director, Dr. Numan and General Amir al Saadi, a personal representative of Hussein Kamil. General al-Saadi listed the key conclusions that a weaponization program would take a decade and require some three hundred Ph.D. specialists and a billion-dollar budget to work in a ten-year program. Dr. Khalid Ibrahim Saeed replied that he could complete the project in less than half of that period—relying on Nuclear Research Centre's potentials without asking for additional funds. At this moment, Jafar looked at both Humam, and Numan amazed. Amir al-Saadi said, "the matter is over. The program has been sent back to the nuclear organization. We wait for your implementation as per the proposal Khalid Ibrahim Saeed."

Dr. Khalid Ibrahim Said, director of the NRC, was promptly put in charge of weaponization. Kamil transferred the group of Khidir's scientists and engineers back to the IAEO in November 1987. The administrative changes did not influence my team's work, mainly because our work was not at the al-Tuwaitha site.

After almost two months, we were visited at our headquarters in the trade unions building by several officials led by IAEO chairmen Dr. Humam, Dr. Ibrahim Saeed, director of NRC, and Dr. Abdul Haleem al-Hajaj. I reviewed with them my designs and work progress in Project 100. Humam appreciated my work and asked me to continue with the same coordination with Saeed regarding anything related to my work. He represented the beneficiary authority of Project 100. The design of Project 100 was completed in one month.

Project 100 took six months to complete. The construction contractor was Saad Public Company (Project 95, Location 1157) under 4D activity supervision.

During this period, I was introduced to some scientific and administrative leaders, including Drs. Abdul Qadir, Abdul Rahman, Hikmet Naeem Jilu, Fadel Abid Muslim, and Numan al-Nuaimi and engineer Dhafir Selbi.

Group 4 (G-4)

The IAEO chairmen formed a new unit, Group 4 at al-Tuwaitha, dedicated to weaponization design. Dr.Saeed was chair, and Dr. Hikmet Jilu was vice-chair.

G-4 included the following activities.

Theoretical study, 4A, chaired by Dr. Mohammed Abdul Zahra
Experimentation, 4B, chaired by Dr. Abdullah Kandush
Metallic uranium, 4C, chaired by Dr. Riyadh al Jarah
Engineering design, 4D, chaired by Dr. Alaa al-Tamimi
Electro-mechanical engineering, 4E, chaired by Dr. Zaghlul Naum
Material experiments 4F, chaired by Dr. Ghazi al Shahri
Other activities and services

The overall cost spent on the construction of project 100 al-Atheer site was approximately one million Iraqi dinars. Construction was completed in May 1988.

Several changes occurred in the organization of G4 after its establishment. In 1989, Activity 4A was split into two new Activities, 40A and 40B. The new Activity 40A was assigned the task of theoretical and computational aspects, whereas, Activity 40B was assigned the task of experimental studies.

The Chemistry Activity was also split into Activity 40 CF, to handle fuel-related work, and Activity 40 CR, to take radio-chemistry work. An Engineering Services department was established before moving to the al-Atheer site.

Later, Activity 4G was split into two activities, 40G and 40N. Late in 1990, Activity 40N was further divided into three activities 40N, the 40S and 40M.

These changes reflected the progress in the understanding of the work and the technical requirements. Due to these changes, the number of teams was also changed.

External Explosion Laboratory

Project 100 was completed in June 1988, but G-4 refused to receive the project before conducting the dependability test. The project was safe for repeated explosions, and vibration did not affect the future experimental findings.

It was agreed with G-4 that the enterprise Qaqa supply the explosive charges. The Building Research Center measured the vibration value, acceleration, and sound intensity in multiple places in and out of Project 100. Meanwhile, Engineer Dhafir Azzawi, al QaQaa' Public Enterprises, who supervised the charge's preparation and explosion, ordered all employees to keep away from the test location.

Yet my designing team and I chose to stay in the lab; we were confident of our calculations, and that confidence transferred to others. Drs Ghazi al-Shahir and Sabah Abdul Noor from G-4 joined us to stay in the lab. al- QaQaa' Public Enterprises prepared the explosive shipments, and the test was carried out twice.

The test proved that the design and execution of the team were exemplary. Project 100 was ready to start its experiments. I felt proud of the excellence of the design of the labs after external and internal explosions. It was an unprecedented achievement in a country like Iraq. I took responsibility for the design, and I lauded the efforts of my design team.

During the test of external explosive lab

Design Authority (5100)

On June 22, 1988, the chair of the IAEO issue an order[29] transferring the activity 4D of G-4 of Office 3000 with the tasks previously assigned to it to Office 5000 and appointed me as the president of Design Authority (Authority 5100). This decision meant that my duty was not limited to G-4 and included all the atomic energy projects.

Functional teams were set up among Authority 5100 activities to prepare plans for the laboratories and installations of Group 4 with the code name of Project 190 out of the al-Tuwaitha site. I was invited to attend a meeting with the IAEO chair. Engineer Khalid Jameel, head of Office 5000, and Engineer Abdul Jabbar al-Zubaidi, chair of Authority 5200,

---

[29] IAEO chair order number 898, June 22, 1988.

were at the gathering and Dr. Khalid Ibrahim Sayed, director of Group 4, and Drs. Hikmet Jilu and Faiz Beriqdar, head of the NCR.

Dr. Humam informed us that we should start work on a new project. The completion of the design of laboratories and installations belonging to Group 4 by Design Authority 5100 were to be transferred to the Execution Authority 5200 immediately.

The project's location would be at farms owned by the organization in Latifiah, on the western side of the Tigris River, thirty-five kilometers south of Baghdad. I expressed my reservation regarding the selection of that site. I knew the nature of the soil in that region and the level of the groundwater there. Our project required dry soil that could withstand big loads. I asked to initiate a thorough study of the location before making any decisions. The chair accepted my request and asked me to submit the survey within a week.

The same day, I returned to the headquarters of the design authority in Karadah al-Sharqiah. I asked some engineers to visit the proposed location in Latifiah and submit a detailed report. The required study suggested changing the site for the following reasons.

The farms in the Latifiah area lay between the Latifiah and Tigris Rivers. The groundwater there was high, and the soil was soft and muddy. The location was close to al-Tuwaitha, separated from there by the Tigris. The construction site could be easily observed by any satellites over the area and arouse suspicions. The place was also close to the Baghdad International Airport, passengers inside the plane at takeoff or landing could take pictures of the project. Agricultural lands would be destroyed by any building there, and anti-aircraft units and rocket launchers surrounding the project would be apparent. In addition, the Latifiah location would be close to residential areas.

I proposed using the same site of Project 100 in Jurf al-Sakhar, sixty-eight kilometers southwest of Baghdad and close to Misayab; it was the new site for laboratories and installations of Group 4 (Project 190). The location was far from residential areas and void of farms, and there were no problems related to the soil and the groundwater level. The chair approved my proposals.

The following day, I visited Project 100 accompanied by some engineers.

Project 190 would be built adjacent to Project 100 after leaving a safe space between the two projects that would be linked by an internal road.

Project 190 was principally involved in materials research and servicing the nuclear weaponization program. The project includes the following material processing production lines:

Casting and purification of metals

(1) Primary casting.
(2) Precision casting
(3) Beryllium casting.
(4) Purification of casts.

Powder metallurgy.

(1) Powder production.
(2) Powder forming.

Hot forming

Administration services buildings and other annex buildings are needed for the industrial complex.

The laboratories and workshop buildings required to house this equipment were approximately eight thousand square meters. It was estimated that another USD 300M was needed for these buildings. An additional cost resulted from the fact that an electrical power substation, water treatment and sewage plants would be required.

In late 1988, Engineer Imad Hamoodi was appointed as Project 190's manager; he came from the Construction Authority (5200).

The Design Authority (5100) started to send completed design drawings to Construction Authority (5200). The purpose was to commence excavations of foundations immediately.

A team was designated by Design Authority to supervise the work on site.

Passive Protection Plans

After I was appointed chair of Authority 5100 on June 22, 1988, I was summoned by the IAEO chair Dr. Hammam Abdul Khaleq, to his office. He assigned me as the advisory body (Formation 1040) on passive protection requirements. He asked me to apply the camouflage measures taken at the al-Tuwaitha site to Project 190 design to protect it.

The al-Tuwaitha Site

Iraq provided appropriate protection to the nuclear reactor site in al-Tuwaitha by installing an air defense system around Baghdad and the nearby reactor.

Surrounded by a sand berm 6.4 kilometers around and thirty meters high, the site contained the French Osirak research reactor that Israel had destroyed in 1981. The purpose for that was to force enemy aircraft to fly high and be subject to Iraqi air defenses. The earthen berm was covered with trees. An irrigation system was set up to irrigate the trees and enhance the green cover over the barricade. The leaves of the trees would absorb a high percentage of the power of isotopes and reduce any radioactive clouds' activity if the reactor exploded or a bomb was dropped on the Russian reactor. Thus, it would alleviate damage in Baghdad and the area around the al-Tuwaitha site.

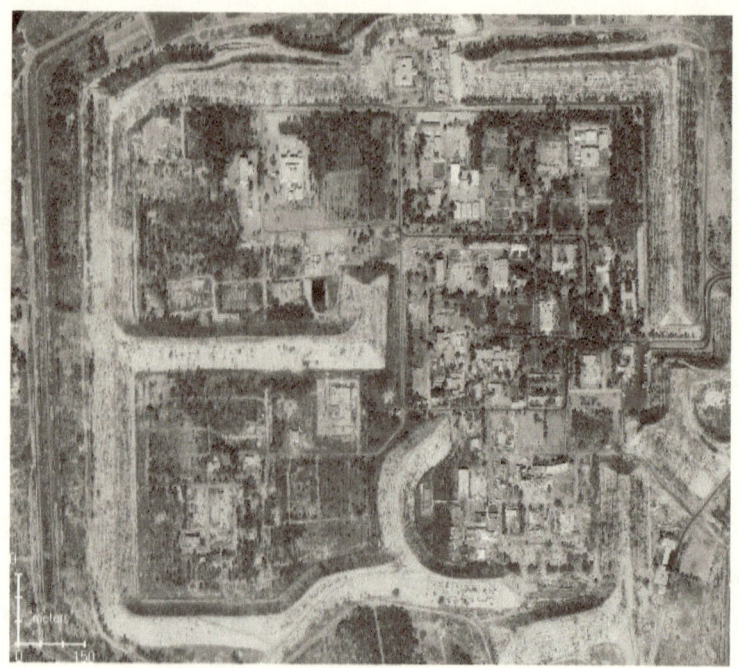

al-Tuwaitha passive protection during the 1980s

The area above the al-Tuwaitha site was protected by balloons filled with hydrogen or helium to impede aircraft flight over the site, particularly at low altitudes.

There were warning stations (radar of various types, human observatories, etc.) distributed. The flight was tasked with monitoring the skies over Baghdad via an air umbrella (interceptor fighter jets) for surveillance starting at dawn and ending at sunset.

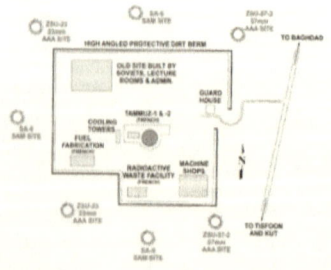

al-Tuwaitha site active protection

Passive protection measures relied on camouflage at the al-Tuwaitha site, the leading site of the Atomic Energy Organization known to all as a nuclear site. The positive protection surrounding the area was relied on by the air defense to repel any attack.

The camouflage networks consisted of two layers of PVC arranged in a certain way. A metal membrane adheres to a smothered network such as a fishing net with an internal tissue layer. It contained circular openings to regulate air leakage from the outside.

Camouflage networks provided optical reconnaissance and nearby infrared, radar, and thermal surveillance. The outer layer of the network reflected light and infrared rays. In contrast, the internal tissue layer scattered and absorbed the bulk of radar waves and distorted them before colliding with the target. The radar waves reflected from the target were similar to the surrounding earth.

Project 190[30]

Passive protection principles adopted at Project 190 include the followings

No foreign consulting or engineering entities were to be used in planning, designing, or implementing the work there.

Site design, buildings, laboratories, and implementation would be handled exclusively by Iraqi engineering expertise.

Applying all camouflage principles during the construction process and during the operation of the project.

Reducing the likelihood of air and satellite reconnaissance identifying the site as a nuclear center.

Reducing the probability of distinguishing and identifying the site tasks as a nuclear center.

Planning and distributing buildings on site with appropriate safety distances to hinder many of them from being damaged by a single airstrike.

Designing the buildings to look like industrial buildings from the air.

Hiding any nuclear radiation activity from buildings.

---

[30] Name changed to al-Atheer site in 1989.

Designing vital facilities to resist the impact of indirect air strikes (250-kilogram bombs) to protect laboratory personnel and equipment.

Developing a plan to rehabilitate buildings and vital points of the project if they were bombed by air to be restarted as soon as possible.

Accepting the possibility of being bombed because the bunker had been built to endure direct hits.

Al-Atheer Center Site Selection

In August 1988, the IAEC decided to construct laboratories and supporting facilities for Group 4. Two sites were considered, the first being around site 100 described in the previous section and the second at Latifiya. The latter was dismissed since it was in a largely agricultural area. Site 100 was selected; since the site was already under development and the existence of infrastructure administration buildings, roads, electrical and water services would expedite the work.

The Design Authority of IAEC designed all premises and site services of the al-Atheer project. No detailed work or studies were made for the process flow of materials within the buildings.

The total area of the site chosen, later named al-Atheer, was 5 km2. The principle of the design of the site was based on zone separation. Each zone was allocated for a specific function (administration zone, laboratories zone, material laboratory buildings zone, etc.).

The center of the site was chosen for the administration zone. The critical buildings were placed far away from the site center to give the impression that the administration buildings were essential. they were of similar external appearance.

The alignment of the buildings and the distances between them were chosen to avoid having two buildings hit by the same aircraft. Window openings were minimized to reduce damage from possible explosions.

It should be noted here that the design of the site and buildings and the design of the services to various processes were accomplished depending entirely upon participating engineers' knowledge and experience. Neither foreign training nor foreign experts were available.

Experience gained from shelters constructed in Iraq and the design of systems adopted in Tamuz reactors and information available in various engineering handbooks, such as industrial ventilation and various literature - listed separately - assisted in designing the internal and external explosion laboratories. Also, the installation instructions provided by the suppliers of some of the main equipment (furnace, isostatic presses, etc.) were helpful.

Planning and Work Program

In Aug. 1988, G4 estimated that the task could be accomplished within a time frame of about three years. A detailed plan was made in which the tasks of G4 were divided into primary and secondary missions, totalling 962. A time estimate for completion was assigned to each task considering the availability of experience and manpower. Technical relations between the various tasks were defined. A computer program "Super Project Plus" was employed to obtain the critical path. PERT charts were obtained and used for discussions and follow-up. Each scientist or engineer responsible for any task was requested to submit a monthly progress report. This information was fed into the computer and followed up by detailed discussions.

CHAPTER 4

Weaponization Moves Forward

Petrochemical Project 3 (PC-3)

In early 1987, Dr. Humam and Dr. Jafar discussed the issue of weaponization. This was a complex challenge in terms of design and engineering. They agreed that it would be unwise to dedicate human resources to another de- manding line of research and development at that stage, as Groups 1 and 2 were still facing many challenges with enrichment technologies. To prepare this report, Dr. Humam asked Dr. Khidir Hamza, who had written one of the earliest proposals for a weapons program in 1971–1972.

In fall 1988, Humam sent a report to the presidency office describing the program's delay in constructing three workshops at the al-Tarmiya site and suggested solutions. al-Aqaba bin Nafi General Establishment had caused the delay. He pointed out that these facilities were dedicated to an enrichment program.

The presidency office referred the report to Hussein Kamil as the responsible minister. The President's office called a meeting of MIC and IAEO to reach a recommendation. Kamil chaired the meeting; he accused Humam of falling behind the schedule of the weapons program. He also claimed that the situation would improve if the MIMI were in charge of the program under his leadership. The president transferred the nuclear weapons program from the IAEO to the MIMI the following week. Kamil maintained Group 1, focusing on gaseous diffusion as a separate entity to stimulate competition.

In January 1989, a Republican decree was issued to create an entity carrying a Petrochemicals Project 3 (PC-3) phony name led by Dr. Jafar. He would be directly linked with the minister. The program was split into two tracks, each focusing on a different enrichment technology. Each was subject to more scrutiny and pressure to deliver. Kamil maintained Group 1, which concentrated on gaseous diffusion as a separate entity to stimulate competition.

PC-3 was much larger than the IAEO, which before PC-3 was formed numbered only about 2,000 people. The size of the combined IAEO and PC-3 was about 7,000 people.

During the previous stage, before establishing PC-3, Jafar was the scientific leader of the nuclear program who provided it with outstanding executive and scientific leadership. He attracted the best Iraqi minds and led them in an advanced and complicated program under terrible pressure caused by Hussein Kamil in his capacity as supervisor of the nuclear program.

New Organization for PC-3

Hussein Kamil put pressure on Dr. Jafar and PC-3 to make rapid progress. He lacked a realistic understanding of timelines and the technical requirements to succeed under his pressure.

After creating PC-3, a new structure and names were adopted. The al-Rabea factory in Baghdad in the Zafaraniyah produced the ion sources. The factory was managed by Engineer Basil Alwan with the assistance of Engineer Samer Kadhim.

The al-Jazira factory near Mosul was designed to produce yellowcake from uranium ore received from Akashat. The site was managed by Dr. Ahmed Shanshal with the assistance of Dr. Abdul Sattar Kazem al-Taie. The al-Safa factory, an electro-magnetic enrichment plant at al-Tarmiya, was managed by Dhafir Silbi. The al-Tuwaith site, which included Group 2, was managed by Dr. Nuaman al-Nuaimi.

The al-Fajr factory near al-Sharqat had been built using Yugoslav design drawings. The site was managed by Dhafir Silbi with the assistance of Engineer Mowafak Aziz. The al-Atheer weaponization facilities were supervised by Dr. Khalid Ibrahim Saied.

The structure included the establishment of the following directorates.

- Design Authority—concerned with preparing detailed designs for new projects and supervision, headed by Dr. Alaa al-Tamimi
- Construction Authority—responsible for the implementation of all construction and services projects, headed by Engineer Abdul Jabbar al-Zubaidi
- Engineering Services Authority—responsible for the installation, operation, and maintenance of services for all sites, managed by Engineer Muammar Mohammed Ali
- Specialized Transport Authority—responsible for specialized transport between sites, headed by Dr. Fadel Abdul Muslim al-Janabi
- Industrial Coordination Authority—responsible for coordinating the sites of PC-3 and all official bodies in Iraq, headed by Dr. Abdul Sattar Kazem al-Taie
- Safety Authority—concerned with industrial and nuclear safety reports headed by Dr. Sami al-Araji
- Authentication Authority, responsible for collecting the scientific, engineering, and administrative reports issued by PC-3, headed by Dr. Imad Yousif Khadori
- Administration Authority—responsible for all administrative, financial, and foreign procurement affairs, headed by Engineer Adel Abdullah Fayyad.

Passive and Positive Protection Advisor

Jaafar issued an order[31]on September 28, 1989, naming me as the advisory body on matters of negative and positive protection of PC-3 facilities and locations; this confirmed the order of the chair of the Atomic Energy Organization (Formation 1040, which had already been added).

All PC-3 sites were under construction, so all the new measurements should match existing conditions.

---

[31] Order (highly confidential) number 715, September 28, 1989.

al-Tarmiya Site

The possibility of developing a passive protection plan on this site was limited because the project was designed and built by a foreign company. Therefore, we had to be satisfied with standard procedures for passive protection and consider it an industrial project unrelated to the nuclear program not to draw attention to its importance.

al-Sharqat Site

The site was still under construction and had almost no equipment during the Second Gulf War in 1991. Under the assumption that the al-Sharqat facility was similar in scale to the al-Tarmiya facility, it could be detectable by ten-meter resolution imagery due to its large scale. However, it differed in the details of its layout. The site was surrounded by a high wall with watchtowers and armed guards.

The passive protection plan was limited based on the fact that a foreign company designed the project. The plan relied on merging facilities with the surrounding environment.

al-Jazira Site

Facility 212 processed U_2O_8 and U_3O_8 into UO_2. The raw U_2O_8 and U_3O_8 were by-products of the Akashat fertilizer plant, not part of the Iraqi nuclear program. Facility 244, adjacent to 212, processed the UO_2 from Facility 212 into uranium tetrachloride (UCl_4).

The development of the passive protection plan was based on camouflage networks; an Iraqi company built the project. The plan merged the facilities with the surrounding environment and deployed camouflaged networks to suit each building.

Zafaraniyah Site

The passive protection plans for this site were in line with those of nearby industrial facilities.

al-Atheer Site

In June 1987, a decision was taken to select a site to construct explosive testing laboratories to include an external and an internal explosion laboratory and related administrative buildings.

The site chosen was in the neighbourhood of existing testing facilities at Jurf Al-Sakher near Al-Mussayib town. The laboratories were given the project code site 100.

The al-Atheer site was involved principally in atomic weapons design; it served the weaponization program and functioned as a materials production and testing center. It was designed and build by Iraqis only. The other atomic projects were designed and built by Italian, French, Brazilian, and Yugoslavian companies.

Al-Atheer covered approximately 35,000 square meters and was about seventy kilometers southwest of Baghdad; it was carefully designed to avoid suspicion. It was discovered only after Operation Desert Storm in October 1991. Project 100, the testing site for hydro-dynamic explosives and a nonfissile natural uranium core in a bunker suitable for nuclear explosion simulation trials, became part of the al-Atheer facility after 1989.

We faced considerable difficulties in designing the material halls laboratories, the carbide building, and the casting building to carry out substance tests, mechanical endurance tests, and neurotic source examinations of the bomb assembly and internal explosion laboratories.

We had to know the technical requirements of each laboratory. I prepared technical questions for the personnel at each facility and building who would work later in those labs (G-4). Most of them were specialists in physics and chemistry.

Everybody was working under pressure. The process required designing an explosive system that contained a detonator consisting of a bridging wire to control the power of shaping the shock waves of pressure explosion.

Work intensified during 1989. Many of the project's buildings and laboratories were constructed. The expansion process led to some premises close to Project 1157 (Military Industrialization and the shorting field), which necessitated the construction of soil barriers 150 meters long and 15 meters high. The base's width was 70 meters and extending along with Building 532 of Project 1157. The purpose was to avoid any overlap between the two projects.

The aim was to transfer new activities affiliated with Group 4 from its headquarters in al-Tuwaitha (the al-Aseel factory) to the new site (al-Atheer Center) in Jurf al Sakhar r. All the designs of the al-Atheer site were completed by the end of 1990.

All essential activities of Group 4 concerned with developing the nuclear weapon in Jurf al Sakhar were called the al-Atheer site in 1989 after the creation of PC-3.

Al-Atheer's main activities focused on trials associated with preparing the uranium core for the bomb, casting, metallurgy, core assembly, explosive lens assembly, and tectonics testing.

Joint work continued between the Design Authority and Group 4. The aim was to determine the requirements of each lab for each activity. Many technical problems needed to be solved. Including how to choose the type and nature of the laboratory, the area of the workspace, kinds of examination, design and distribution of air conditioners, and the variety of filters and purifiers whose styles, techniques, and volumes differed according to the type of work done in every lab. The purpose was to discharge the air from the rooms and guarantee that it would not be contaminated.

Coding of projects at Al-Atheer site works

(1) IAEC decided to choose the al-Atheer site for G4 projects on 26-May-1988. Hence, it was given the code "8526".

(2) A different code (site 190) was given for construction teams and contractors. Since almost all works were of construction type at that stage, the code (190) became more common than 8526.

(3) Construction and civil work projects were coded by four digits as follows:

(3.1) 1st digit (6000) represented the code for the nuclear research center at Tuwaitha (office 6000), where G4 was originated.

(3.2) The 2nd digit represented the technical specialty of the building as follows:

100 - Explosive Labs 200 - Electronics
300 - Analytical Chemistry 400 - Physics

500 - Material Science

600 - Physics with Explosives 700 - Administration

800 - Services for material building 900 - General Site Services

(3.3) The third digit, in most cases, represented the main task of the building.

(3.4) The fourth digit represented the main labs in the building.

The goal of the al-Atheer plant program was to design and manufacture; it was composed of the following principal parts.

production of polonium-210 nuclear initiator (polonium 210 metal beryllium)

production of natural uranium metal (core enriched uranium metal)

reflector natural uranium metal

temper-hardened iron

explosive lenses

electronics systems (triggering, control, and guidance)

For camouflage from aerial reconnaissance, all buildings had identical external shapes. My main goal was to make sure that the al-Atheer premises bore no distinctive elements. The exterior walls, corridors, and installation of the electro-mechanical systems were designed to endure indirect airstrikes. The design details paid attention to each lab's furnishing, lighting, ventilation, and safety precautions. Facilities are designed so that work is not interrupted during the site's exposure to air raids to ensure the availability of water, electricity, maintenance during the airstrike.

After completing the first phase, most Group 4 staff moved from their duty post in al-Tuwaitha to the al-Atheer facility. The equipment, machinery, and testing system were moved, installed, and put into operation. The minister of industry and military industrialization opened the facility of this phase on May 7, 1990.

Kamil, the minister of Industry and Military Industrialization, visited the al-Atheer center in May 1990, accompanied by Jafar. The minister was received by Group 4 director Dr. Khalid and Dr. Alaa al-Tamimi, chair of the Design Authority. During the visit, the minister had time to see the lab of external explosions (Project 100). He asked me about the possibility of observing the project by satellite. I replied that the ability of army satellites was not known to me. However, I assured him that the lab's exterior design did nothing to indicate it was involved in a nuclear project.

Opening al-Atheer on May 7, 1990

Even in May 1990, the site was not ready as essential utilities, including electricity, water, and suitable air systems, was not in place.

By June 1990, G-4 scientists had begun conducting experiments on implosion-type nuclear bombs with highly enriched uranium cores surrounded by shaped charges.

The main buildings designed and built at al-Atheer were metallurgy laboratories (Project 6530), polymer laboratories (Project 6250B), carbide building (Project 430), powder technology laboratory (Project 6250A), and the internal explosion test laboratory (Project 6000), which was associated with the control building. The site also included manufacturing, maintenance, electrical, and welding workshops and administration facilities.

The most important buildings at the site dealt with uranium metal at high levels of purity, melting, pouring, reshaping it mechanically, casting to carry out substance tests, mechanical endurance tests, and examinations of the bomb assembly.

Some buildings were completed in the second half of 1990, and equipment that had arrived by that time was installed. However, some essential services were still incomplete, such as the electric supply, water system, compressed air and special ventilation. After Aug 2, 1990, and due to the embargo, the equipment ceased to arrive. This equipment represented the major part of the supply.

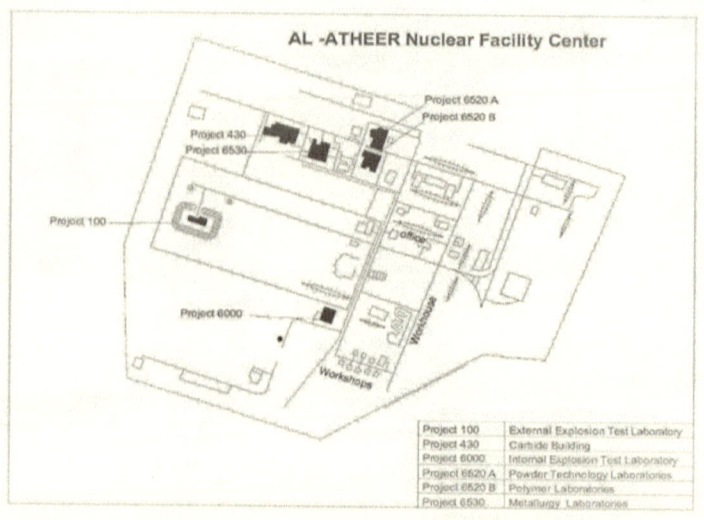

al-Atheer's most important buildings

Most G-4 staff moved from the al-Tuwaitha site to al-Atheer in Jurf al Sakhar after completing Project 190, the second phase of the al-Atheer facility, at the end of 1990.

Kuwait Invasion and War

All the engineering and mechanic activities to build a nuclear device were almost in their final stages at the al-Atheer facility. Saddam, who was supposed to be fond of power, would never jeopardize such a program by any means.

On August 2, 1990, Saddam invaded Kuwait; he desecrated its land, vandalized its facilities, looted its shops, and blew up oil wells there, provoking the whole world to stand up to him. The invasion wasted the importance of Arab unity, destroyed national ties, and led to an unprecedented division in the Arab ranks.

The invasion was shocking news to all. The situation was open to all possibilities. Weariness overwhelmed the rational people who recognized the seriousness of the problem. The official media launched a campaign to cheer the invasion as the military operation was very swift.

UN Sanctions

On August 2, the UN Security Council held an emergency meeting and adopted UN Resolution 660. They condemned the invasion and demanded that Iraq withdraw its forces immediately and unconditionally. On August 6, Resolution 661 reaffirmed Resolution 660 and noted Iraq's refusal to comply. The council took steps to implement it in Iraq under chapter 7 of the United Nations charter.

However, on August 8, Iraq announced its annexation of Kuwait. Weariness gradually intensified due to Iraq's failure to respond to the Security Council resolution to withdraw from Kuwait. Hundreds of thousands of US troops and their allies were being mobilized. The Security Council adopted twelve resolutions between August 2, 1990, and November 29, 1990; all but the first were adopted as per chapter VII of the United Nations Charter. According to this chapter VII, the Security Council shall determine the existence of any threat to the peace, breach of the peace, or act of aggression and shall make recommendations or decide what measures shall be taken to maintain or restore international peace and security.

These resolutions were 665, 664, 662, 661, 660, 678, 677, 674, 670, 669, 667, and 666. Most of these resolutions were concerned with the Security Council's preoccupation with weapons.

Resolution 678 gave Iraq a last chance to implement the resolutions above, including unconditional withdrawal provided for in Resolution 660; otherwise, member states would have the right to enforce it. This resolution was a sort of legal authorization to launch the war. The resolution was supported by twelve member states, opposed by two—Cuba and Yemen—while China abstained from voting.

Emergency Plan

By the time Iraq invaded Kuwait, the nuclear project still lacked an indigenous source of fissile material; its enrichment plants were still far from producing highly enriched uranium (HEU). Kamil visited al-Tuwaitha accompanied by LG Amir al-Saddi, vice-chair of the MIA. He chaired a meeting attended by Drs. Jafar and Saeed and gave the order to start a crash program immediately. The plan was to extract HEU from the fuel to enrich its portion further and build a nuclear weapon within six months.

On the same day, Dr. Jafar issued an order to extract HEU from fresh fuel used for the Soviet IRT-5000 reactor and the French radiant fuel. The project was to be completed in three months. The use of nuclear fuel for objectives other than powering an atomic reactor would mean that Iraq had violated the Nuclear Arms Prevention Treaty (NPT), which prevented to use of nuclear fuel for military purposes, which Iraq had signed in 1972.

The crash program required an urgent execution of the installations and laboratories of G-4 in Jurf al Sakhar (Project 190). There was an urgent need to finalize the vital projects at the al-Atheer site so that the latter could be ready to receive enriched uranium due on November 18, 1990.

Most of the G-4 departments relocated to al-Atheer; three departments remained at al-Tuwaitha.

Construction Delays

There were many delays in construction, which took at least a year, due to the late arrival of equipment and poor management. A series of measures were taken in late spring 1990 to make up for the lost time. Each lagging task became subject to intense review, and more pressure was put on those responsible for constructing the laboratories and the assembly hall where the nuclear bomb would be built.

On August 2, 1990, the invasion of Kuwait and the adoption of the Security Council's resolutions had severely affected complete construction at the al-Atheer site. The UN sanctions blocked Iraq from receiving any equipment from international companies.

To accelerate work at the al-Atheer site (Project 190), Dr. Jafar summoned me to expedite the completion of the installations there. He said that the minister's authority would assign me as the project's executive director to push the work until completing the project's vital structures.[32] It was exceptional for a senior director to be given authority to complete the task of the minister's management. It was a severe responsibility I was asked to take on; I felt grave personal consequences if I failed.

Becoming the project's executive director with ministerial authorization and its significance motivated me to complete the final vital projects as soon as possible. I opened a site office at Jurf l Sakhar and held meetings with the technicians, engineers, and administrators. I motivated them and solicited their opinions on how to solve problems and save time. I also held meetings with contractors and their engineers and technicians urging them to accelerate their work on the project. I told everybody that we would work day and night if the provisions were made available. All the cadres executing the project were asked to be present at the site continuously. Contractors were mainly concerned with payment, so I created a mechanism to pay them within twenty-four hours. I acted immediately to determine the critical path for completing the vital installations. They were projects 6510, 6520, 6530, and 430.

Considerable efforts were taken to complete the polymer lab, the materials lab buildings, the internal explosion test laboratory and associated control building, the manufacturing workshop, the maintenance workshop, and three warehouses.

The necessary measures were taken to specify the needed equipment. I obtained Jafar's approval to transfer the electro-mechanical machines at the al-Fajar factory to the al-Atheer site. Electrical boards were manufactured at the Dijlah factory in Zafarahniya. Electrical cables were obtained in local markets; those markets were filled with items looted from Kuwaiti. Cranes were prepared at Nasir Public Enterprises in Taji, near Baghdad.

All other services were finished. The works included administrative buildings, a restaurant, roads, a water station on the Euphrates River, and

---

[32] Order number 1060 on November 20, 1990 issued by the Ministry of Industry and Military Manufacturing.

a secondary power station. We were completing the vital enterprises in January 1990.

Table 4.1

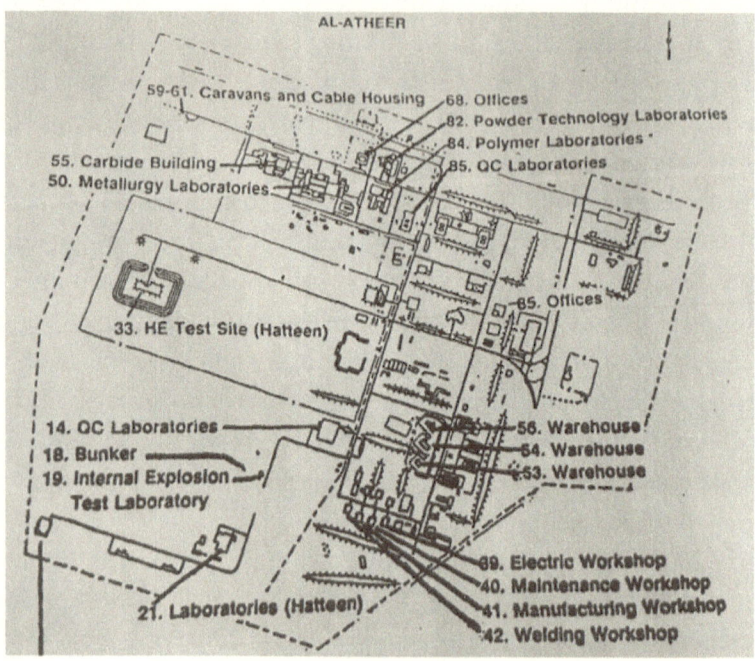

al-Atheer Facility site in 1991

The following is the name of the al-Atheer Facility site in 1991

6110	Soil Berm
6200-1	Electronic Labs
6200-2	Electronic Labs
6300	Analytical Chemistry
6400	Physics Labs
6500	Code for material buildings.
6510	Original code for U-casting
6520	U-Melting and casting building
6520A	Administration for 6520
6530	Powder Technology building.
6560	Material Physics building.

6570	Characterization of Materials building.
6580	U-Chemistry & Polymer building
6600	Internal Explosion Lab.
6610	Gas Gun Lab.
6620-6650	Codes were not used
6660	Internal nuclear initiator
6700	Administrative building.
6705	Housing For Security People
6710	Laundry building.
6715	Restaurant
6720	Telephone Exchange building.
6725	Control & Security building. "Internal Reception"
6726	Shelters
6730	Car Park & Maintenance
6735	Engineering Services Work Shop
6740	Medical Center
6745	Stores
6755	Fire Brigade
6800	Code was not used
6900	Roads & Fences
6950	Water Projects
6951	Sewage System
6952	Fire Fighting network
6953	Irrigation network
6954	Industrial water network
6955	Drinking Water System
6970	Electrical System
6975	Permanent Electrical Station
6979	Internal Electrical
6980	Electrical Network Systems (H.T., L.T., Lighting)

Most G-4 members moved from al-Tuwaitha to al-Atheer in Jurf al-Sakhar after completing Project 190 on the due date and of the required quality.

The crash project was stopped by the outbreak of the war in January 1991.

Security Council resolutions provided an opportunity for a permanent military presence of the great foreign powers in the region.

President Bush of the US gave Iraq an ultimatum ending at midnight on January 15–16, 1991, to withdraw from Kuwait. When Iraq refused, the US-led Arab-International coalition launched its war. At 2:30 a.m. on January 17, 1991, I was awakened by strong explosions. I used a caravan as a mobile home in the water project station set up on the Euphrates River fifteen kilometers east of the al-Atheer project. I was accompanied by my wife and four-year-old son Hussein. I could not leave them alone at my home in Baghdad as I felt that war would soon begin.

The scene was terrible. I stood by the river. Dozens of jet fighters flew over my head on their way to Baghdad. I saw many antiaircraft rockets heading toward their targets but not hitting them due to their altitude. I watched several military-industrial establishments on the other side of the river being bombed and consumed by fire.

It seemed that the sky was on fire. The noise of the jet fighters and rockets shook our automobile. I was terrified, but I pretended to be courageous for the sake of my terrified wife and son, who were overwhelmed by the sights and sounds. I was alone on the road. The night was illuminated by shaky moonlight concentrated on the nearby military enterprises.

I decided to go to the holy city of Kerbela; we arrived there at five in the morning. We found thousands of Iraqis walking in the courtyard of Imams al-Abbas and al-Hussein (grandsons of Prophet Mohammed). I felt reassured. I went to the house of a friend, who welcomed me. I noticed that he had received many friends, some of whom had come from Baghdad seeking safety. At nine that morning, I left my family there and went back to my work site at Jurf Alsakhar, hoping to return in the evening.

I arrived at the al-Atheer site before noon. I found the manager of my office, Yahya Suhail, there. He told me that the project had not been hit the previous night. Few other people were at the site. Everybody was concealing their fear of more air raids. Work was stopped on Jan.17, 1991.

I left in the afternoon for Kerbela accompanied by Shakir, my driver. I told him I would spend the night in Kerbela. I was planning to take my family back to Baghdad. A primary school in the district of Zafaraniyah was selected to be a substitute office during the war.

Air raids throughout Iraq continued from January 17, 1991, till the cease-fire on February 27, 1991. The jet fighters and guided missiles were bombing camps, airports, bridges, water and power stations, telecom systems, ports, government premises, stores, and even civilian shelters.

On January 21, 1991, during a meeting with Dr. Jafar, I was asked to visit al-Tuwaitha and report the damage inflicted on each installation due to the air bombing. I paid several visits to the al-Tuwaitha site during January 23–25, 1991. I met Dr. Humam, head of the AEO, at the site to inspect the damage inflicted on the Tammuz"(July) nuclear reactor. While we were there, the air defense at the location launched smoke bombs to conceal the site from the planes attempting to photograph and thus assess the airstrikes' damage. The Gulf War bombardment damaged al-Tuwaitha very severely.

During the second week of the war, the presidency office formed a committee for air raid shelters. I was nominated as a member. I was astonished that the state planned to construct air raid shelters that required years to build while we could not find a place to hold meetings other than a secondary school building.

Bombing Nuclear Projects

When the Second Gulf War began in 1991, no one knew where Iraq's nuclear program had arrived, neither Israel, America, Iran or others. Iraq's nuclear program has remained secret. The Iraqis have largely succeeded. They are skilled in concealments, and I admire their tactics.

Those were the words of David Kay.[33]

The new Iraqi nuclear projects built after 1981 were distributed to five sites across Iraq: al- Tarmiya, al-Sharqat, an industrial complex in Zafaraniyah, al-Jazira, and al-Atheer. The passive protection measures that had been taken proved to be successful. Satellites and aircraft surveillance had not identified any of the sites as nuclear sites. They had not been included in the objectives of the US airstrike campaign. They thus survived

---

[33] The chief US inspector of the IAEA Commission on Iraq's weapons of mass destruction. In a special interview with al-Wasat Journal on April 20, 1992.

severe damage from the airstrikes, especially during the first three weeks of the war.

The primary uranium enrichment site at al-Tarmiya (Safaa site) was not among the targets of the US bombing campaign. However, US intelligence had identified the site as related to the industrial military but linked to a missile program.

Nuclear sites were targeted during aerial bombardment as follows.

al-Tuwaitha Site

This site declared to the IAEA was attacked by Iran in 1980 and Israel in 1981. The site bombed violently from January 20–21, 1991, till the cease-fire on February 28.

US warplanes bombed al-Tuwaitha, destroying the July 14 research reactor and the July 2 reactor building, including administrative buildings, and destroying thirty significant buildings.

The NRC at al-Tuwaitha, with approximately one hundred structures in the compound, had been familiar to the Americans at least since Israeli F-16s had attacked it in 1981. American F-16s struck it during the first week of the air bombing, and F 117s visited the compound on February 18, 19, and 23. On February 23, thirteen F-117s bombed al-Tuwaitha in good weather. At least eighteen of twenty-six bombs hit structures in the compound.

The US attack almost led to a major nuclear disaster. Buildings contained a wide range of atomic fuel rods containing highly and medium-enriched uranium. If that fuel had leaked into the groundwater would cause an environmental and humanitarian disaster, the world would not have been able to rid Iraq of its long-lasting effects. If the poison had gotten into the air, it would have been dangerous for Iraq and neighboring countries.

al-Tarmiya Site

In 1992, an American magazine[34] published a report quoting a pilot who had bombed the al-Tarmiya site. The pilot said that he was flying

---

[34] IEEE Spectrum, April 1992, 29:4.

back to his base after February 15 and had three remaining bombs. He noticed large buildings near the Tigris River and decided to target the three most significant buildings. After dropping the bombs, he filmed the targets he had attacked. The pilot observed intense and unexpected human activity around a facility, signifying the[35] importance of the site. Risking Iraqis being nearby in these dangerous conditions, aircraft returned to the location on February 16 and bombed the site again. On February 19, the area was attacked again and wholly destroyed.

al-Sharqat Site

The twin site of the Tarmiya project, 250 kilometers north of Baghdad in the al-Sharqat area (al-Fajr project), was heavily bombed on February 23. I assume that comparing similar aerial images of the al-Sharqat and Tarmiya sites led to the decision to attack the al-Sharqat site.

al-Atheer Site

Camouflaging the al-Atheer site from aerial surveillance was one of the most significant challenges I faced during the design of the al-Atheer center during its construction phase (1987 to 1990). I planned to ensure that the site's passive protection succeeded. I must mention the creative engineer Essam, who distributed essential buildings under the passive protection plan discussed earlier. The al-Atheer facilities were not subjected to aerial bombardment, which lasted more than forty-two days and did not indicate essential targets.

In the second week of the war, Project 100 was bombed directly, causing the fortified building to move slightly, but it did not suffer any other damage. However, it was likely to[36] be targeted as part of the passive protection plan for the al-Atheer site, as previously stated.

The last night before the cease-fire, the restaurant building and the power plant outside the project were targeted.

---

[35] "Seeking nuclear safeguards. How Iraq reverse engineered the bomb." IEEE Spectrum, April 1992, 29:4.

[36] See chapter 3, where project targeting was a hundred percent likely under the project's negative protection plan.

As mentioned before, a unique philosophy was adopted for the passive protection of Project 190 (renamed al-Atheer after 1989). Passive protection proved successful because the site was not hit by airstrikes during the Second Gulf War and was discovered only through documents and by chance seven months after the cease-fire.

al-Jazira Site

On February 20, al-Qaim, a uranium extraction facility, was attacked. More than twenty-five strikes attacked the al-Jazira facility during the war's last two weeks. The critical areas of the 212 facility processes were not damaged. Facility 244 adjacent to 212 processed the UO_2 from facility 212 into uranium tetrachloride (UCl_4); it suffered no more than 20 percent aggregate damage.

Industrial Complex in Zafaraniyah

The industrial complex in Zafaraniyah survived the US bombing in 1991. The two factories were highly technological and promised by IAEA inspectors to be an "elite Iraqi industry." The complex was then bombed by US aircraft on December 17, 1998. Operation Desert Fox lasted for four days.

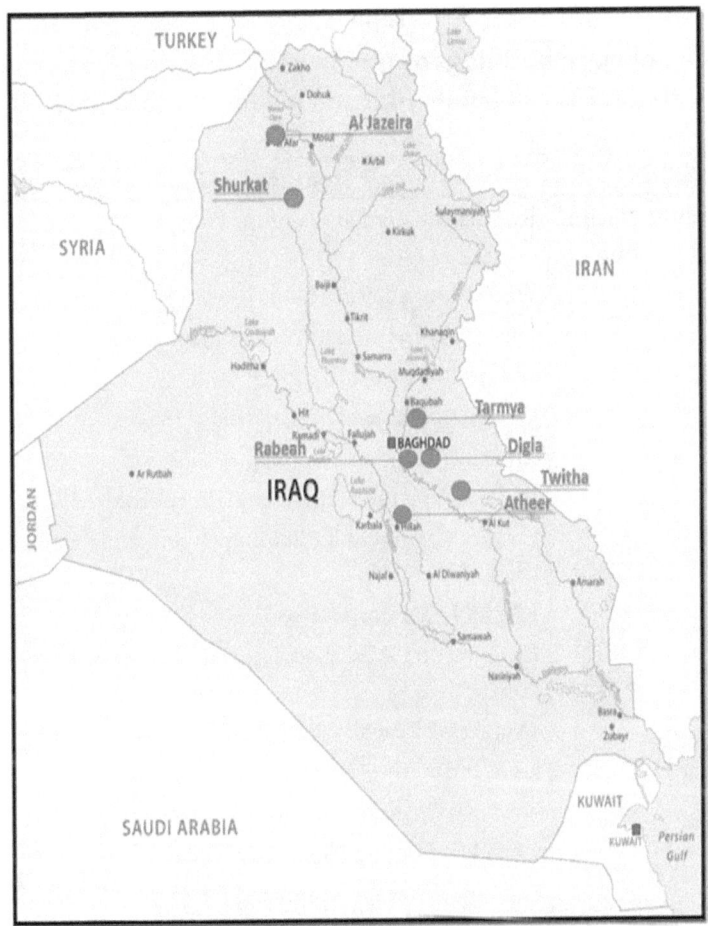

Map of nuclear locations

Table 4.2
List of main buildings of Iraq Nuclear Program destroyed during the aerial bombardment (January - February 1991)

Site	Destroyed Buildings
al-Tuwaitha Nuclear Research Centre	Radiochemistry Laboratories (Bldg. 9)
	Physics Department (Bldg. 10B)
	Laboratory for uranium Metal Preparation (Bldg. 10)1
	IRT-5000 Reactor (Bldg. 13)
	Computer Hall and Offices (Bldg. 13 part)
	Electrical Sub-Stations (Bldg. 14, 72, 84)
	Radioisotope Production Department (Bldg. 15A)1
	Quality Control of Radioisotope Production Department (Bldg. 15B)1
	LAMA Laboratories (Reprocessing), Bldg. 22)
	Experimental Workshop, laser and Plasma Studies (Bldg. 23)1
	Tammuz-2 Reactor (Bldg. 24)
	Store and workshop (Bldg. 28)
	Decontamination Laboratory (Bldg. 27)
	Chemical Coating Laboratory (Bldg. 30)
	Cooling Tower for Tammuz-2 Reactor (Bldg. 31)
	Radioactive Waste Treatment Station (RWTS, Bldg. 35)
	Calibration Laboratories and Decontamination Area (Bldg. 41)
	Laboratories for Material Processing (Bldg. 83)
	Laboratories for Uranium Treatment and Liquid Radioactive Waste (Bldg.64)
	Laboratories for Experimental Physics and Measurements (Bldg. 68)
	Hydrogen Station (Bldg. 70)
	Sewage Station for 30 July Project (Bldg. 71)
	Experimental Research Laboratories for Fuel Fabrication (Bldg. 73 complexes)1
	Cooling Tower of bldg. 80 (Bldg. 79)

	Laboratories for EMIS Development (Bldg. 80)1
	Laboratories of UCl$_4$ Preparation and Purification (Bldg. 85)1,2
	Chemical Enrichment Laboratories (Bldg. 90)
	Mechanical workshop (Bldg. 57)
	Material studies (Bldg. 63)
	Electrical engineering design labs. (Bldg. 82)
al-Atheer Centre	High Explosives Test Bunker and stores (Bldg. 33)2
	Offices of Activity 40B (bldg. 79)
	Electrical Laboratories (Bldg. 94)
al-Tarmiya EMIS Facility	EMIS stage 1 Separator Building (Bldg. 33)
	Air Conditioning Units (Bldgs. 21-23, 94-96, 244, 248)
	EMIS stage 2 Separator Building (Bldg. 245)
	Electrical power Sub-Stations (Bldgs. 5, 38, B1, 243, 228)2
	EMIS Separator Wash Room (Bldg. 225)2
	Waste Treatment Building (Bldg. 216)
	Chemical process building (Bldg. 210)
	Chemical process building (Bldg. 230)
al-Sharqat EMIS Facility	EMIS Washing and Cleaning (Bldg. C-034)
	EMIS Washing (C-054)
	Electrical Power Supply (Bldgs. B-029, B-027, B-020, B-032)2
	Utility Building (Bldg. B-031)
	Cooling Towers (Bldg. B-033)
	Equipment Hall (Bldg. B-051)
	Main Power Station (B-046)
	EMIS stage 1 Separator Hall (B-021)
	Workshop (B-003)
al - Qaim Uranium Purification Facility	Uranium Purification Building (Bldg. 300)
al-Jazira Uranium Processing Plant	UO2 Production Plant (Bldg. 000)
	UCl4 Production Plant (Bldg. 400)

	UCl4 Production Plant Utilities
	UO2 Production Plant Utilities

Iraq further levelled the building to the ground.

Building further destroyed under IAEA supervision.

Bombing al-Amiriya Civilian Shelter

Earlier in the 1980s, twenty-five shelters were constructed in Baghdad and its suburbs by a Finnish company. In 1985, more such places were built, including their isolation from the electro-magnetic sphere generated by a nuclear strike. The Americans concluded that these shelters were dedicated to Iraqi leadership and bombed these sites without paying attention to innocent civilians.

As a family, we all slept in windowless internal rooms on the ground floor to avoid glass shattering if a bomb fell near our house.

On the night of February 13, 1991, two F-117s dropped laser-guided two-hundred-pound bombs on the al-Amiriya shelter containing four hundred men, women, and children. My house, which was about seven hundred meters from the shelter, trembled terribly. The first rocket penetrated concrete walls and hit the shelter's roof, causing its steel doors to be closed. A second rocket was fired and turned the shelter into an oven that melted the families' bodies who had fled there for protection. US jet fighters turned the shelter into an oven. Only nine survived the strike.

Shelter for the Minister's Palace

During the fifth week of the war, I was notified at about 7:00 pm that Hussein Kamil wanted to meet me at the oil ministry premises at 10 am the next day. I was worried about this call, especially since Kamil had assumed the minister of defense's position and the supervisor of the Ministries of Industry and Military Industrialization. The next day, I was at the premises of the Ministry of Oil. He ordered officers to escort LG Amir Alsaadi to his office in the secretariat room right after he arrived.

Two hours later, Kamil walked out of his room, saw me, and ordered me to follow him while running to his Mercedes. He was quiet and polite as he drove us through Baghdad, which was almost deserted, at 3:00 p.m. as the sun would set in two hours and the terrifying airstrikes would recommence.

I thought of the mission he wanted me to take on. I was confident that it would be related to the war. He drove over the al-Jumhuriya Bridge, and we entered the Karadah district, the presidential palace area. We headed toward the suspension bridge and then to Karadah al-Sharqia

After crossing the suspension bridge, he drove right and went over the river dam. Several kilometers later, he stopped. He got out and walked toward a site that was in the early stages of construction. He informed me that the project was being built on land the president had granted to his sons, Odei and Qusai. Since the project was still in its early stages, he planned to construct a shelter that would be a refuge for its inhabitants and be secluded from the external world. He pointed out that the shelter would not be helpful in this battle. Still, Iraq was expecting many battles ahead as long as it did not comply with the "imperialist" schemes as he called them.

I was astounded by his statement. I pitied the developing nation whose leader thought of building a shelter under his palace. In contrast, Iraq was passing through grave days.

Cease-Fire

The US-led armies entered the city of Kuwait on February 26, 1991. President Bush declared a cease-fire from one side as of February 27, 1991, after the lapse of a hundred hours from the land war's start. When the war came to its end, the Iraqi army had been devastated, and its government had been significantly weakened. All observers expected that President Saddam's regime would be toppled. Bush indirectly urged Iraqis to revolt against the government; he announced that the coalition forces' main task was to liberate Kuwait. Iraq's political regime change was an internal affair.

Large-scale displeasure began to be felt among the ranks of the withdrawing Iraqi army. In March 1991, a popular uprising in Iraq started when an unknown soldier directed his tank gun at a picture of Saddam in Basra's central square. The event was the first spark of the revolt that overwhelmed the southern parts of Iraq, followed by cities in the north. However, units of the Republican Guards and some army units that remained loyal to Saddam managed to quickly put out the fire of revolt. Millions of Kurds in northern Iraq began to leave the country, heading to Turkey and Iran.

The liberation of Kuwait has accompanied the destruction of Iraq at all levels and created a new reality in Iraq.

Reconstruction Campaign

Throughout the forty-two-day war, Iraq had been subjected to bombardment by more than 100,000 tons of explosives, including hundreds of tons of depleted uranium ammunition. The campaign of bombing caused the death of 70,000–100,000 Iraqi troops. The US coalition lost 505 soldiers, including 472 US soldiers. About 40,000 soldiers were wounded. On the other hand, 30,000 soldiers became prisoners of war. Iraq lost thousands of tanks and artillery and hundreds of jet fighters. Air defense and telecommunication stations, missile launchers, military research centers, and warships were destroyed. The US airstrikes destroyed infrastructure, including schools, institutes, universities, telecom stations, radio and TV stations, oil refineries, seaports, bridges, railways, power stations, and water supply installations.

Two weeks after the official cease-fire, the Iraq government commenced the reconstruction campaign under the slogan Fie on the Impossible.

The Iraqi nuclear project staff managed to reconstruct electricity, oil, and industries in record time. Breaking the monotony and bureaucracy of work and cooperation among different ministries and departments contributed to elevating obstacles and facilitating efforts. The spirit of patriotism was a vital motive to achieve something. When participants of the reconstruction campaign asked about what they were doing and why, the answer was straightforward: to provide our countrymen and women with things that would make lives easier, including fuel, electricity, roads, and buildings. Political differences had no place. The result was fascinating. The Iraqi cadres were creative in performing tasks that were not achieved during the pre-Kuwait Iraq.

The job of rebuilding the power supply sector was assigned to Dr. Jafar. To this end, the scientific and engineering staff of PC-3 was mobilized to carry out this mission, especially since most of the nuclear project activities had been suspended during the airstrikes.

Reorganization

In April 1991, a decision was made to dismantle the Ministry of Industry and Military Manufacturing into two entities: the Ministry of Industry and Minerals and the MIC, and the decision kept Hussein Kamel as their supervisor in addition to being the minister of defense and supervisor of the Ministry of Oil. On July 8, 1991, Dr. Jafar was appointed director of the MIC. Kamel wanted to enhance the MIC staff of PC-3 in the reconstruction phase.

The Design Authority and Construction Authority became a new establishment under the establishment of industrial execution. I was nominated as general director of this new establishment. The industrial execution establishment was tasked with constructing the damaged buildings at the al-Dura power station, Nasiriyah power station, al-Hartha power station, al-Musayib thermal Station, the Taji power plant, and a steam and gas molasses power plant.

CHAPTER 5

Weapons Search and Destroy Inspections

At the end of the 1990–1991 Gulf War, the UN passed Security Council Resolution 687 on April 3, 1991, setting the terms for the cease-fire between Iraq and the US-led coalition. Section C of the resolution called for eliminating Iraq's weapons of mass destruction and some ballistic missiles and established the United Nations Special Commission (UNSCOM).

UN Security Council Resolution 686 adopted on March 2, 1991, reaffirmed Resolutions 660, 661, 662, 644, 665, 666, 667, 669, 670, 674, 677 and 678 (all in 1990), suspended military activities against Iraq, and mandated that all twelve resolutions would continue to be in full force and effect. The resolution forced Iraq to release all Kuwaiti citizens and other detainees, hand back Kuwaiti properties, pay war damages, and appoint military offices from the US-led coalition to monitor the cease-fire.

The Kuwait war's main result was the destruction of Iraq's infrastructure and its army, which was regarded as one of the regions mightiest. Iraq was strictly isolated under UN sanctions that suffocated the country's economy for thirteen years.

The UN Security Council resolutions froze large amounts of Iraqi assets in international banks. Five percent of Iraq's revenues would pay compensation, estimated at $52 billion, to those affected by the invasion— about a hundred countries and international organizations primarily Kuwait.

After Iraq accepted UN Security Council Resolution 687 (the so-called cease-fire resolution) on 6 April 1991,

Saddam Hussein accepting inspection was terrible by all the standards of a respectable leader, and he had to step down.

Weapons of Mass Destruction (WMDs)

After the cease-fire, Iraq had entered the inspection phase for weapons of mass destruction. The UN Security Council adopted Resolution 687 on April 3, 1991.

In the 1980s, the IAEA was performing regular safeguards inspections in Iraq. The agency was implementing a safeguard system that depended on Iraqi declarations of nuclear material and nuclear activities. After the Gulf War, it discovered that Iraq had violated the treaty by implementing a covert program to produce and weaponize fissile materials. Resolution 687 was a cease-fire agreement, not a safeguards agreement. The resolution gave the IAEA responsibility for activities that were different from regular safeguards activities.

The resolution stipulated that a commission would search for and destroy any Iraqi WMDs. Paragraph 12 of the resolution mentioned above stated,

Iraq shall unconditionally agree not to acquire or develop nuclear weapons or nuclear-weapon usable material or any systems or components or any research, development, support, or manufacturing facilities related to the above.

Iraq was driven out of Kuwait by force. The nuclear program verification process became the main reason for UN sanctions. The inspection and verification process was very complicated and challenging because of the US's interest in keeping Iraq sanctioned and isolated.

IAEA Nuclear Inspections

Resolution 687 provided that Iraq would place all its materials meant to produce nuclear weapons under the exclusive control of the IAEA. The destruction, removal, or rendering harmless as appropriate of all items was specified. On completing the destruction of Iraq's WMDs and whatsoever was related to it, Iraq would have to declare it within fifteen days as of the resolution date. When the Security Council was satisfied with the results, the sanctions provided in Resolution 661 of 1990 would be cancelled.

In implementing Resolution 687, Dr. Jafar, director of PC-3 projects, prepared a memorandum that listed all centers and institutions working in the nuclear weapons program and their activities. It would submit all these details to Kamil, chair of the Military Industrialization Authority. Kamil issued his orders to uncover the activities and location of al-Tuwaitha only. He ordered the preparation of alternative programs to be submitted to the UN team if they requested inspecting any place other than al-Tuwaitha. Other nuclear locations were not to be mentioned as per his orders. These locations included al-Tarmiya, al-Shurgat, al-Jazira, al-Atheer, al-Rabie mechanical factory, and Dijlah electrical factory.

The IAEC seniors could not get a coordinator with the IAEA because the Iraqi authority did not want to put its senior scientists in contact with the inspection teams. The nuclear materials flow calculations (input-output) were a fundamental conflict issue with the IAEA until PC-3 chief Jafar settled the problem with the inspection teams.

His instructions stipulated that Kamil substances of general use must be separated from those related to the nuclear program. Equipment associated with the nuclear program should be evacuated by trucks owned by the Republican Guards to new locations where they could be stored or destroyed.

Iraq submitted to the Security Council on April 18, 1991, a first declaration in which it admitted that it owned substances used in manufacturing nuclear weapons. On April 27, Iraq submitted a second declaration acknowledging that it had some nuclear substances and premises in addition to those already known by the IAEA. Furthermore, Iraq declared that it did not possess nuclear weapons or the material used for making them.

The known information about Iraq was limited to that acquired from the IAEA's twice-yearly inspections of the al-Tuwaitha nuclear research center facilities. The center was declared under the safeguards agreement concluded between Iraq and the IAEA in connection with the Nuclear Non-Proliferation Treaty (NPT). Iraq had been a party to that treaty since 1969. These facilities were primarily the IRT-5000 research reactor, the Tamuz-2 research reactor, a small fuel fabrication laboratory, and a storage facility.

Obstacle Resolution 687

Instead, declaring the nuclear program as a whole as per Resolution 687 allowed the cancellation of the sanctions when the Security Council was satisfied. Iraq committed a fatal mistake by not revealing its nuclear program in full. Iraq thought that the sanctions would be lifted once the inspectors failed to find anything.

It was impossible to hide Iraq's substantial nuclear program; it would be uncovered sooner or later. An endless succession of hasty instructions, confusion, and maneuvering failed to bluff the inspection teams.

It seems that the factors listed below had pushed Iraq to decide not to disclose the information to the UN.

To preserve valuable resources by undermining the destruction, removal, and deactivation of certain types and substances associated with the prohibited programs.

To maintain its ability to carry out some prohibited projects after the departure of the UN inspectors.

There was some wariness that other nations might seek to collect intelligence data about Iraq within the framework of UN inspections.

The other critical mistake was the destruction of equipment and machinery without keeping records of that. It was a procedure that violated Resolution 687, which stipulated that the process of destruction should be performed under international supervision.

The consequences of those wrong decisions kept the sanctions effective for thirteen years. The United States exploited it to keep the embargo on; that paved the way for Iraq's invasion and occupation in 2003.

We must ask why Iraq had continued to hide the existence of nuclear programs. The answer may be to let others believe that it had such deadly weapons to deter attacks. Unfortunately, the US has drummed and booby-trapped this same belief despite its proven knowledge of wrong.

The US planned to appear before the world to confront a rogue state armed with weapons of mass destruction that threatened the region's security and world peace until Iraq finally decided to declare that it did not possess weapons of mass destruction. This declaration came too late. Even countries against the war and the international bodies in which neutrality was supposed to be accepted found it challenging to admit it. It was to disguise Iraq's dealings with inspection committees.

UN Search Inspection Missions

The IAEA rejected Iraq's first declaration because it didn't contain anything about important nuclear material in Iraq. On April 27, Iraq declared that it did have HEU and admitted that it had buildings and facilities at al-Tuwaitha other than those visited by safeguard inspectors. The IAEA carried out an immediate on-site inspection of Iraq's nuclear capabilities based on the Iraq declaration and developed a plan to destroy, remove, or render harmless these capabilities in forty-five days.

IAEA director-general Hans Blix set up an action team on April 15, 1991. The rest of the group was made up of three senior professionals and two clerical staff. Their task was to plan and lead the first and all subsequent nuclear inspections.

IAEA sent twenty-seven inspection teams between September 1991 and August 1995. These teams destroyed all the systems, projects, equipment, and items involved in the uranium enrichment program. The first of these inspections focused mainly on the sites and locations Iraq had declared.

The UN inspection team affiliated with IAEA began its visit to al-Tuwaitha on May 15. The first inspectors' mission was to verify the validity of the declaration submitted by Iraq to the Security Council on April 18, 1991. The inspection team confirmed that Iraq did not possess nuclear substances fit for the manufacture of nuclear weapons. After the departure of the delegation on May 23, two buildings at al-Tuwaitha were demolished.

The UN's first team of inspectors visited the al-Tarmiya location and saw the area as a factory that produced high-tension transformers. The inspectors were convinced of this scenario. They reported that the al-Tarmiya site was not nuclear and had nothing to do with nuclear activity.

By the end of its mission, the first team had successfully established a path for future on-site inspections.

The second team of IAEA inspectors arrived on June 22, 1991 and asked to carry out a surprise inspection. It moved to western Baghdad, to a camp for the Republican Guards in the Abu Ghraib district. The base contained some equipment involved in the nuclear program. Kamil had ordered it to be delivered there. The soldiers prevented the inspectors from entering the camp. The soldiers fired their guns to warn the inspectors against any attempt to enter the site. Once the international inspectors had left, the equipment was spirited away at night. The inspectors were permitted access to the site the following day and did not find anything.

It was evident that the inspectors relied on satellite pictures. The camp had nothing to do with the nuclear program. Therefore, it was easy for them to locate where the equipment, including vast magnets of almost six meters in diameter and a weight of more than sixty tones, had been moved. Such equipment was available at the al-Tarmiya site. They were transported on huge trucks but were intercepted by UN inspectors between al-Tarmiya and Baghdad. The cat and mouse game was not equivalent. The inspectors had intelligence and were supported by the Security Council. On the other side, there were improvised and hasty orders. As such, the inspectors had the upper hand, and Iraq was in a weak position.

In the wake of that event, a high-ranking UN delegation chaired by the assistant secretary-general for disarmament affairs arrived in Iraq. The delegation asked Iraq to facilitate immediate access for the agency inspectors to any site they wished to visit. A new Security Council Resolution 707 was adopted. It stipulated that the inspection of any location whatsoever shall not be hampered or delayed.

After determining that it was not possible for Kamil to hide the equipment, President Saddam issued two decrees. The first created a committee whose function was to manage the relationship between Iraq and the IAEA. It would be chaired by Tariq Aziz, the deputy prime minister. That committee put an end to Kamil's role in this field. The second was to destroy all the debris of the mass destruction weapons program from one side by the personnel of the Republican Guards.

On July 7, 1991, Iraq submitted a third declaration of new details about its nuclear program to the Security Council. The letter pointed out the details of uranium enrichment using centrifugation, gas penetration, and electro-magnetic processing. The declaration did not include the armament program (al-Atheer Center) and ignored the al-Rabie mechanical factory and the Dijlah electric factory affiliated with the nuclear program. The third, fourth, and fifth inspection teams focused mainly on verifying atomic substances and enrichments. The inspection teams managed to access all sites without any problem.

IAEA inspectors visited several places, including al-Atheer but did not find any direct evidence of weaponization. However, they certified that al-Atheer and two neighboring facilities constituted a complete and sufficient potential nuclear weapons laboratory and production facility. This combined facility was so big and well-equipped that it could do much more than the limited non-weapons activities the Iraqis claimed to be its purpose. Iraq submitted a third declaration on July 7, 1991, that disclosed clandestine centrifuge and electro-magnetic isotope separation (EMIS) uranium enrichment programs. It maintained that Iraq had complied with the NPT and the IAEA safeguards agreement. Iraqi confirmed no political decision to reveal the weapon program and no formal nuclear weapons program.

Iraq submitted a fourth declaration on July 28, 1991, that listed nuclear materials not previously declared to the IAEA.

The Sixth UN Inspection Team

The Design Authority had moved in 1988 from Karada al-Sharkiah to a building close to the headquarters of the MIA. Many articles and books were mixed up between the Design Authority and other PC-3 engineering departments in the al-Nakabat building.

At 6:00 a.m. on Monday, September 23, 1991, the sixth UN inspection team led by David Kaye arrived at the Design Authority building. The group was divided into two sections. The first went to the basement, and the second accessed the ground floor. In the basement, the inspectors focused on four metal boxes contained microchips of the blueprints of the engineering designs and reports of undeclared activities of the fourth group—mainly concerned with creating the nuclear weapon. It also had documents of PC-3. They also found a report on the progress of the work at the al-Atheer project.

Kaye asked to take the four boxes. He was informed that he could inspect their contents at the center, but the UN team refused. One of the inspectors claimed a kidney problem. An ambulance was called to take him to a hospital. Later, it turned out that his kidney problem had been faked. He had hidden a work progress report on the armament activities in his clothes.

The Iraqi team accompanying the UN inspection group did not allow them to leave. The UN team warned the Iraqi authorities that the notice would end at 4:00 p.m. on that day. The Iraqi authorities were not to hinder the progress of the inspection team's work. They warned the Iraqis that they would notify the UN that Iraq had violated Resolution 687. Jafar came to the building and asked to inventory the documents. The UN inspection team could have a look at the documents he added.

The argument between Jafar and Kaye went on for half an hour. They agreed that they should leave because the reports were to be handed over to them at midnight. The reason was to let the Iraqi team take stock of the items. The inspection team left at 7:00 p.m.

It was evident that the arrival of the sixth inspection team to the Design Authority building and its attempts to inspect specific locations on the premises represented a security violation. The inspection process was performed according to intelligence information.[37]

On September 24, 1991, al-Khairat building's inspectors departed the administration offices of PC-3. As mentioned earlier, the building was a temporary location for PC-3 staff after the nuclear research center's destruction in al-Tuwaitha. The international inspectors obtained detailed data about the personnel and the project's nature by photocopying the files on the premise's computers.

The documents showed that nuclear weapons work had been done at the al-Atheer facility. It also tied al-Atheer with the IAEC and PC-3. The inspectors concluded from documented links between PC-3, the IAEC, and the MIMI that PC-3 was the code name for Iraq's nuclear weapons program. The link was confirmed by personnel files found at PC-3 headquarters.

With the help of PC-3 employment records, several sites were identified as part of the program. It was concluded that the nuclear weapons program was under the general control of MIMI with specific power assigned to

---

[37] Robert Gallucci, assistant head of the sixth inspection team, declared later that a person from the CIA had informed him in August 1991 about the Iraqi nuclear program's location and existing papers on the atomic armament program.

PC-3. PC-3 employee lists showed that Jafar was a senior administrator for the program. Similar documentation showed that he was personally linked to the uranium enrichment program.

I called for an inquiry at the premises of the Military Industrialization Authority on Palestine Street. People called for the detention and investigation of many, including Saeed, Adil Fayadh, and Abdul Sattar al-Taie.

It had been "an administrative error" under such circumstances in Iraq. General Amir Rashid merely recommended that everyone involved with the documents be reduced a rank; he released the detainees. The contents of the G-4 weaponization documents were transferred to optical file disks only when the inspectors were already in the country. The hard copies of the papers were then burned.

Table 5.1

List of main buildings of the sites of Iraq Nuclear Program that survived the war but were later destroyed under IAEA supervision

Destruction Date	Site	Destroyed Buildings	Destruction Method
April- May 1992, IAEA-11/12	al-Atheer Centre	- Carbide (uranium machining), Bldg. 55	Demolition with explosives. Bldg. 33 was filled with concrete and scrap metal; the protective berm has been removed
		- Casting (uranium metallurgy), Bldg. 50	
		- Quality Control. Bldg. 19	
		- Explosion Chamber, Bldg. 1B (cutting with torches)	
		- High Explosives Test Bunker, Bldg. 33	
		- Physics (gas gun), Bldg. 21	
		- Polymer (uranium metal processing), Bldg. 84	
		- Powder Preparation, Bldg. 82	
July -September 1992, IAEA-13/14	Tarmiya EMIS Facility	- Electrical Sub-Stations, Bldg. 5, 38, 243 - EMIS stage 2 Separator Building, Bldg. 245	Demolition with explosives/heavy machinery
		- Electrical Sub-Stations, Bldg. 5, 38, 243 - EMIS stage 2 Separator Building, Bldg. 245	
July -September 1992, IAEA-19/14	al-Sharqat EMIS Facility	- Electrical Sub-Stations, Bldgs. B- 20,B-27, B-29	Demolition with explosives/heavy machinery
		- EMIS stage 2 Separator Building, Bldg. B-21	
November 1993, IAEA-22	Abu Skhair Mine	Abu Skhair uranium mine	Backfilled, shaft door welded and sealed

Note: Electrical power supplies to the Tarmiya and al-Sharqat sites were reduced by order of magnitude.

All known facilities for the industrial-scale production of pure uranium compounds suitable for fuel fabrication or isotopic enrichment have been destroyed, along with their principal equipment.

All known single-use equipment used in the research and development of enrichment technologies has been destroyed, removed or rendered harmless. All known dual-use equipment used in the research and development of enrichment technologies is subjected to ongoing monitoring and verification.

All known facilities and equipment for uranium enrichment through EMIS technologies have been destroyed and their principal equipment.

IAEA inspections have revealed no indications that Iraq's indigenous plutonium production reactor plans proceeded beyond a feasibility study.

The facility used for research and development of irradiated fuel reprocessing technology was destroyed in the bombardment of Tuwaitha, and the process-dedicated equipment has been destroyed or rendered harmless.

The principal building of the Al-Atheer nuclear weapons development and production plant has been destroyed. All known purpose-specific equipment has been destroyed, removed, or rendered harmless.

The entire inventory of research reactor fuel was verified and accounted for by the IAEA and maintained under IAEA custody until it was removed from Iraq".

IAEO Head of PC-3

Dr.Jafar was dismissed from his MIA position as director of PC-3 and was appointed the presidential office's scientific adviser. He asked to continue the task of reconstruction and rehabilitation of the electricity sector. Dr. Humam Abdul Khaliq was assigned to manage PC-3 in addition to his job as chair of the IAEO

The leaked document contained a report on the work progress at al-Atheer from January 1 to May 31, 1990. It was mainly related to the issue of nuclear armament. It is to be noted that the establishments of al-Atheer site, specifically the internal and external explosion labs (Project 100), had been completed and were handed over to G-4 in the second half of 1989.

Inspectors from the IAEA and other intelligence agencies were not acquainted with al-Atheer's significance; it had not been linked to the Iraqi nuclear program and thus had survived the airstrikes. Yet, the document found in the Design Authority's building was the first thread that led to the location's discovery.

The seventh inspection team concentrated on the destruction of the equipment for uranium enrichment. Its members asked questions regarding the development of nuclear weapons. They relied on the information they had gained from the sixth team, which had been headed by Kaye. The report provided information about G-4 at al-Atheer, which referred to the nuclear bomb's design activities.

On October 31, the last day of the inspectors' visit, they submitted questions to Iraqi authorities concerning the design studies, including those concerning the design of hydrodynamic tests.

On October 21, Iraq acknowledged that research and studies had been underway in the area of nuclear weaponization. Iraq admitted that the al-Atheer site had been built to serve the weaponization program and its use as a materials production site.

I was summoned to attend one of the meetings with the seventh international team held at MIA headquarter. The inspectors' questions focused on Project 100. I had signed the design blueprints of the two labs they had found. I assured them that the designs were pure Iraqi without any foreign aid.

On October 11, the UN Security Council adopted Resolution 715 approving the IAEA's plan for ongoing monitoring of Iraq's compliance with Resolutions 687 and 707.

The eighth inspection team visited al-Atheer and thoroughly examined the polymer, internal and external explosion test laboratories, and the control premises. Completing the visit, the eighth team had inspected all sites involved in uranium enrichment or atomic armament. The nuclear program had become fully exposed.

In October 1991, the destruction of the particular sections of al-Atheer center and its interconnected systems required eight tons of explosives. I was there representing the al-Atheer center along with the Iraqi nuclear team. This site signified the greatness of the Iraqi mentality.

Iraq provided its fifth declaration on November 20, 1991, containing information relevant to Monitoring and Verification OMV. Iraq provided a supplementary statement on January 13–14, 1992, but the declaration remained incomplete.

Iraq provided its sixth declaration on December 11, 1991, regarding nuclear programs, information required for OMV under UNSC Resolution 715.

Iraq provided its seventh declaration on March 12, 1992, and handed over a full, final, and complete disclosure (FFCD) to the IAEA director-general. The FFCD consolidated previous statements and treated them as a draft in light of agency questions about their adequacy.

UN Destruction Missions

International inspectors saw al-Atheer as a counterpart of the Los Alamos labs, which had played a significant role in the Manhattan Project. Pop Kelley, the armament expert, described al-Atheer as a center that could be considered a regional hub for examining substances. It had enjoyed a high level of accuracy in design and structure. After weeks of objecting to destroying the al-Atheer Center, labs were damaged on April 14, 1992.

Iraq unwillingly allowed the destruction of al-Atheer's establishment after the IAEA revealed two documents that disclosed plans to develop a nuclear explosive device there.

After discussion, the decision was made that buildings involved in the program and the equipment used or essential for the weapons program had to be taken out or destroyed. The decision to destroy al-Atheer's main buildings meant that Iraq would lack a site to fabricate nuclear weapons.

The external explosion laboratory had been targeted during the early days of the Second Gulf War. A bomb had penetrated it, but the inside of the bunker was intact; explosives could not be used to destroy it. I suggested to the inspectors that the optimal way of doing that was to fill it up with concrete. They went through that process, and we poured several tons of concrete until the bunker was filled.

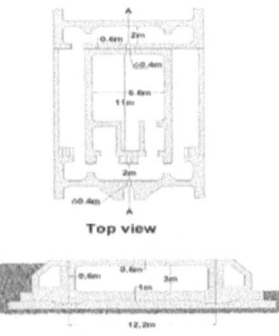

Top view

External explosion laboratory bunker

They used explosives to bring down all the facilities involved in the program at al-Atheer. As well, equipment used or essential for the weapons program had to be taken out or destroyed.

Items destroyed included a uranium metallurgy facility and a third of a tungsten carbide production building. They destroyed dual-use equipment such as special furnaces, precise measuring equipment, metal-coating equipment, machine tools, and isostatic presses.

During 1991 and 1998, fifty-two inspectors were consecutively on inspection, verification, and monitoring duty tours in Iraq. In all, they conducted 299 visits to sites and facilities. UN inspectors solved the story of Iraq's past weapons of mass destruction programs and destroyed any remaining components and infrastructure.

Thus, we reach the beginning of the end of the story of Iraq's nuclear project, which took its last breath in 1992. The IAEA continued to send inspection teams until August 1995. They completed the destruction of all systems, equipment, uranium enrichment, and weaponization program materials.

Administrative Decisions

In October 1991, Hussein Kamel, Saddam Hussein's son-in-law, resigned to the president from his position as head of the Ministry of Industry and Military Industrialization

Saddam accepted the resignation and appointed his cousin, Ali Hassan al-Majid, Kamel's uncle.

In late February 1992, things returned to normal between Kamel and his uncle, President Saddam. Kamel was reinstated to his position as the second man in the regime.

He obtained an order from President Saddam on March 13, 1992, to transfer PC-3 with all its equipment and facilities to the Military Industrialization Commission. The transfer of PC-3 employees to the MIC was a shock to most PC-3 employees.

Kamel, the supervisor of the MIC, issued an order to merge the design authority with the implementation authority. He established an organizational unit on behalf of the General Facility for Industrial Projects (GFIP) with a moral personality and financial and administrative independence. He appointed me as a general director of the new facility on April 1, 1992.

GFIP was assigned to evaluate and restore several electrical power plants affected by the bombing during the war. GFIP and al-Fao company were given a project to build new annexes to the presidential palace, severely damaged during war airstrikes. GFIP implemented the eastern wing as a contest between the two sides (duration and cost).

The idea of an engineering project competition was new to me, and I was not fond of it. Determining the shortest duration for any project depends on what is known about the critical path method. Completing any project in less than that calculated in the critical path meant that the project was not executed under engineering specifications.

My new boss was MG Nizar al-Qusayr, an engineer awarded the rank of honorary general. He issued orders in a way that did not pertain to engineering work other than what prevailed in my previous experience at IAEA or PC-3. At the Atomic Energy Organization or PC-3, I used to discuss any subject related to my work at length until a mutual opinion was reached, not by issuing orders as was the method used in the military. Al-Qusayr did not accept the idea of discussing any issue. He told me that the Atomic Energy Organization's work method differed from the Military Industrial Commission's.

Dr. Humam, chair of the AEO, and Dr. Jafar, director of PC-3, were at the high end of engagement with their employees.

Leaving the Nuclear Project

Starting the week after being appointed director-general of the GFIP, I tried every way I knew to withdraw from my position and return to the faculty and scientific research. I asked General Amir al-Obeidi to relieve me of my mission and allow me to return to the university after the cancellation of the nuclear project. Many other engineers could have served as GFIP's general director. I pleaded with him to present the matter to Kamel and obtain approval to move me out of the MIC. I received ministerial approval to leave my position on September 3, 1992.

I returned to my field of university teaching and scientific research. I joined the teaching staff at the Military Engineering College and the Faculty of Engineering at the University of Mustansiria. I was there for four years until I left Iraq in 1996.

Hussein Kamel Defects

On August 8, 1995, two sons-in-law of Saddam Hussein defected to Jordan. This was a dramatic blow to the Iraqi president. LG Hussein Kamel Hassan, minister of Military Industrialization and the Industry and Minerals Ministry, had arrived in Amman along with his brother, LC Saddam Kamel Hassan. The latter had been in charge of Saddam's guards. Both men were married to daughters of Saddam and had sought diplomatic asylum in Jordan; that was quickly granted.

Iraq summons the head of the International Inspection Committees, Rolf Ekeus.[38] Ekeus provided many previously unauthorized documents on various weapons programs. The Iraqi authorities took Ekeus and his team to Hussein Kamel's farm and handed him containers full of important documents. On August 20, 1995, Iraq handed over documents to the UNSCOM and IAEA with information that had allegedly been withheld on Hussein Kamel's orders without the knowledge of the Iraqi government.

---

[38] Between 1991 and 1997, he was the director of the United Nations Special Commission on Iraq, the United Nations disarmament observers in Iraq after the Gulf War. In late July 2002, he reportedly said in the *Svenska Dagbladet* newspaper that during his time in his position, he attempted to resist attempts by the US to use the commission to perform espionage.

The Haider House Farm cache consisted of more than 500,000 pages of documents.

Leaving Iraq

In 1995, I decided to leave Iraq after learning of colleagues' travel who had previously worked on the nuclear program. The travel ban had been lifted for all atomic energy workers. This was perhaps one of the conditions imposed by the UN to ensure that Iraq did not return to weapons programs before lifting the economic embargo.

My wife and son were granted new passports, but my name remained on the travel ban list. I realized that I couldn't travel outside Iraq, which gave me distress and boredom.

Fate entered here in my favor. In life, some incidents occur that may seem trivial but become very important. I had contracted gout, a type of arthritis caused by crystal salts in joint tissue. This illness directly relates to the liver function responsible for expelling or keeping salts in the body. The treating doctor recommended that I have an MRI (Magnetic Resonance Image) scan test to determine the nature of a tumour found on my liver as soon as possible. I had to leave Iraq as MRIs (Magnetic Resonance Image) were not available locally.

I submitted a request to the Special Medical Committee to leave the country to get the MRI, and I received permission from it to do so. Still, the passport department told me that any exceptions to the travel ban had to be granted by the Intelligence Service, not the Special Medical Committee.

I was racing against time as I didn't know the nature of the tumor in my liver. I turned to the former head of the Iraqi Atomic Energy Organization, Dr. Jafar, with whom I had a working and friendship relationship.

Dr. Jafar, vice president of the IAEC, is a creative scientist and a great humanitarian figure. His late father, Dr. Dhia Jafar, served as Iraq's minister of the economy during the monarchy in the 1940s. I had the privilege of working directly with him from 1987 to 1992. I served as head of the Design Authority and then director-general of the GFIP of the IAEO.

Jafar was then the undersecretary of the Ministry of Industry. He asked me to submit a new application with all medical reports and the support of the medical committee to travel outside Iraq for treatment. A week later, his office manager called to tell me that the passport department would allow me to travel out of the country but only one time.

I was leaving my precious homeland in the summer of 1996. I had done the impossible to leave the land of a hard life. As I approached the Jordanian border in a rental car with my wife and son (who was only nine at the time), my anxiety flared up. I wondered if, indeed, I could leave Iraq and create a new life. I asked if my family and I would ever return to Iraq. Oh my God—How wishes can turn into reverse desires at any moment. My heart began to pound. Driving over the border into Jordan was a fantastic experience. I saw a bright future for the three of us.

During the first months of my arrival in Amman, I underwent the required medical examinations and learned that the tumor on my liver was benign; it was probably due to an old blow my liver area had suffered. I had to do other tests three and six months later to compare with the current medical results. I calmed down a little after the results of the reassuring tests and decided to wait in Amman for the following tests.

Subsequent tests confirmed that the tumor was not serious and that there was no need for additional tests. That made me comfortable and energetic enough to complete my dream of providing a new life for my family and me. I stayed in Amman until we left as immigrants in early 1998.

Currently Accurate Complete Documentation (CAFCD)

United Nations Security Council Resolution 1441 was adopted unanimously by the United Nations Security Council on November 8, 2002. it was offering Iraq under Saddam Hussein "a final opportunity to comply with its disarmament obligations" that had been set out in several previous resolutions (660, 661, 678, 686, 687, 688, 707, 715, 986, and 1284). Iraq agreed to the Resolution on November 13, 2002.

The Iraqi government submitted detailed reports according to Resolution 1441 on nuclear, biological, chemical, and missile weapons programs on December 7, 2002. Still, the United States of America and

Britain started the war on March 19, 2003, without authorization from the UN Security Council.

Iraq Invasion

The uncertainty that Iraq was free of weapons of mass destruction was due in part to the behavior of the Iraqi regime.

Iraq believed that the suspicion of its possessing deterrent weapons would protect it regionally and internationally and give it the appearance of strength and prestige.

Doubts about Iraq's possession of weapons of mass destruction before the invasion were not limited to the Americans. Even members of the UN Security Council had such doubts, which were clearly expressed by its Resolution 1441 adopted unanimously on November 8, 2002.

The resolution offered Iraq under Saddam Hussein "a final opportunity to comply with its disarmament obligations." It justified what has subsequently termed the US invasion of Iraq.

On December 7, 2002, Iraq submitted a significant report to the UN Security Council about its military and civilian programs that could have military applications about four months before the invasion of Iraq.

The Iraqi report answered questions that remained unresolved after the inspections that had taken place in Iraq between 1991 and 1998. Iraq provided evidence to confirm that it had disposed of all nuclear, missiles, chemical and biological programs.

The report contained about 20,000 pages; its main sections, in Arabic and English, were subsequently translated into French, Russian, and Chinese, the languages of the permanent members of the UN Security Council.

Since then, no one has spoken about the authenticity of these reports submitted by Iraq to the United Nations.

Although the inspectors did not get any positive results on the claim of weapons of mass destruction, on March 19, 2003, President Bush Jr. announced that the USA and coalition forces had begun military action against Iraq.

Appendix A contains the original text of A summary of the Currently

Accurate Complete Documentation (CAFCD)report on the nuclear weapons program submitted to the UN Security Council on December 7, 2002.

Return Home

Changes in our lives are often due to a story or a tale, even when things happen suddenly. Early in 2004, a website that dealt with Iraqi affairs announced that Baghdad's city council was searching for a mayor for the city. I thought that this position's sensitivity required the candidate to be a member of the ruling political parties. Still, the truth soon revealed that the Coalition Provisional Authority decided to encourage local governments to take democratic steps and elect officials democratically, a shift Iraqis had never experienced before.

On April 20, 2004, the interim coalition press release stated that the election process had been successful after weeks of intensive negotiations between Baghdad's council members. They courageously established nomination rules and guidelines for voting to determine the mayoral selection procedures. Despite the problems faced by some regions of Iraq, the mayor's choice showed unequivocally the consensus that had overcome divisions in the council. The Baghdad city council announced that I, Dr. Alaa al-Tamimi, had been elected mayor of Baghdad.

On May 23, 2004, the Baghdad city council organized a ceremony at the conference palace to swear in Baghdad's first mayor elected without government intervention or party support in Iraq's history.

After 18 months as a mayor, I was forced to leave Iraq due to a threat to my life. The religious militia occupied my office (Mayor Office) on August 8th, 2005.

CHAPTER 6

Evidence and Implications

I attempted my best in this book to avoid becoming entangled in political matters. Drawing valuable lessons from experience begin with an accurate record of what happened. I tried to focus as much as possible on my roles throughout my career in the nuclear project, 1987–1992. I did not want to take on the position of a political analyst.

I have sought to provide in this book a true and accurate Iraq Nuclear program not published elsewhere. Then I tried to explain why Iraq failed to acquire nuclear weapons and how state visions worked as variables influencing the performance of nuclear weapon programs such as foreign assistance and the intervention of the political leadership's interest in the micromanagement of the nuclear weapon program.

Iraq regime decisions through the president Saddam or his son of law Hussein Kamel were important reason variable shaping the performance and the failure outcome of the program through their choice of management strategies, and centralized control.

However, we have to agree that Iraq had the right to obtain nuclear technology for peaceful purposes according to the nuclear non-proliferation treaty. The bombardment of the Tammuz (July) atomic reactor in 1981 by Israel drove the Iraqi decision to violate the treaty without thinking of its severe consequences. Saddam believed that obtaining the bomb would allow Iraq to deter attacks by its two main enemies, Iran in the east and Israel in the west. [39]

---

[39] "We have to have this protection for Iraqi citizens so they will not be disappointed and held hostage by the scientific advancement taking place in Iran or in the Zionist entity," Saddam said in 1981. "Without such deterrence, the Arab nation will continue to be threatened by the Zionist entity and Iraq will remain threatened by the Zionist entity." Woods, Palkki, and Stout, *A Survey of Saddam's Audio Files*, 266.

Iraq was involved in a clandestine nuclear weapons program. It was the first time an NPT state had violated the treaty by implementing a clandestine program to produce and weaponize fissile materials.

President Saddam had acted unjustifiably and illogically by occupying Kuwait. He put Iraq under the microscope of friendly and hostile states leading to tragedies that Iraq has been suffering from for decades to come.

It is essential to point out that the efforts of the nuclear staff will be a source of pride for each Iraqi citizen who contributed to that excellent scientific and engineering effort. The experience revealed a determination to realize an unprecedented strategic and technological accomplishment by Iraq confronted by enemies that sought to destroy it under the pretext of removing weapons of mass destruction.

After the Second Gulf War, everything changed due to UN resolutions. Resolution 687 was to deal with three verification tasks—to determine the nuclear program's scope and extent, to destroy, remove, or render harmless the components of that program, and to start implementing an OMV system.

Why Iraq Failed to Develop a Nuclear Bomb

Iraq came close to creating nuclear weapons. I summarize here three vital practical factors behind the failure of Iraq's nuclear program.

President Saddam's Priorities

Saddam Hussein did not put nuclear weapons at the top of his priorities. His lack of focus on this goal was a significant reason Iraq did not reach it. Any country that begins a project to build an atomic bomb must stick to it until the program ends. Saddam invaded Kuwait in 1990. The invasion attracted the world's attention when Iraq's nuclear program had achieved remarkable development at a particularly sensitive phase of the atomic program. Otherwise, Iraq would have been highly likely to acquire nuclear weapons by the mid-1990s.

Should Saddam have been more patient with preserving Iraq's progress and development programs? He had ordered the beginning of the nuclear program, but then he created conditions for its destruction after the invasion of Kuwait. Iraq agreed to the terms of the international coalition after the Second Gulf War. Concerns about Iraq's economic troubles drove it to face economic sanctions and inspections after the 1991 Gulf War. As a result, the nuclear weapons program was permanently dismantled.

Confusion's Consequences

Nuclear programs were not micromanaged or closely monitored. Saddam had no realistic idea of how long it would take to acquire nuclear weapons. He pressured his scientific leaders to produce. It was better to work quietly, but Kamel continued to rush, urge, and demand progress. At the same time, nuclear project officials dealt with time factors and political and international changes.

Dr. Khadouri, in his book,[40] wrote,

Dr. Jafar asked him to index a box full of sensitive reports and important correspondence and store it on a CD. Khadouri allowed himself to read some of these correspondences and did not rest for its contents, which contained deliberate exaggerations or ambitious conclusions hoped to be achieved. Hussein Kamel signed these reports to Saddam Hussein in 1990.

The Program Scientific Leadership

In April 1985, Saddam summoned the leaders and five senior members of the IAEC, Chairman Izzat Ibrahim, Vice-Chairman Humam, Commissioners al-Kittal, Said, al-Mallah, Jafar, and Selbi, to review the progress of the weapons program. They were taken to the Radwanya district of Baghdad, where he was waiting in a caravan. As the meeting started, Humam reviewed the ongoing projects. Dr. Humam completed his presentation by announcing that the program would fulfil its objective to reach a new milestone in the programs within five years, by 1990. Saddam reacted with visible emotion, according to the recollection of Selbi. When Saddam invited the other commissioners to comment on Dr. Humam's presentation, they remained silent. No one said anything.

---

[40] Dr. Imad Khadouri, Iraqi Nuclear Weapons Squadron, Arab House of Sciences, Beirut, 2003.

After the meeting, the commissioners were presented with new cars and mobile caravans.

It seemed out of character, as no one in the nuclear establishment had previously set a timeline or deadline.

Dr.Humam's promise, in my view, about a new milestone in the enrichment programs and not deliver the bomb. The commitment was remarkable as the notice was made while enrichment technologies were being explored at a theoretical or experimental stage, and no significant breakthroughs had yet been achieved.

On another occasion, when they received the order to begin the crash program, the scientists did not tell Hussein Kamil that it would be impossible to succeed within the given deadline. As Iraq stood on the precipice of the 1991 Gulf War, Kamil and Saddam may have believed that a nuclear weapon was within reach.

On the other hand, Kamil did not always understand the consequences of his decisions. He lacked technical know-how, which the scientists occasionally exploited by obscuring their reports and recommendations in overly technical language.

The Pride of Iraq

I want to pay tribute to the scientists, engineers, technicians, and everyone who worked on the Iraqi nuclear project. This experience will remain a point of pride for every Iraqi who contributed to this outstanding scientific and engineering effort and advance Iraq's scientific capabilities.

The project revealed a sincere desire and determination behind an unprecedented Iraqi scientific achievement met by opposition from Iraq's enemies. Ultimately, Iraq was destroyed through the pretext of eliminating weapons of mass destruction. The Commission on the Intelligence Capabilities of the United States about Weapons of Mass Destruction. Testified on March 31, 2005, that the intelligence community was "dead wrong" in assessing Iraq's weapons of mass destruction capabilities before the US invasion.

This book is my attempt to lift the veil on the Iraq National Nuclear Program that has witnessed a wonderful expression of the Iraqi mind.

I hope this book will renew the Iraqi memory and inspire the national spirit of Iraqi youth, who want to restore their country after the destruction of the occupation and political sectarianism.

A phoenix obtains new life by rising from the ashes of its predecessor. Every time Iraq falls, invaders and occupiers burn its cities and books to destroy its civilization, a greater Iraq rises. This has happened many times. The Mongols occupied it, as did the Persians, Turks, British, and finally the Americans, but Iraq reemerged and became more influential on regional and international levels every time.

APPENDIX

After completing the draft book and sending the manuscript to the publisher, I received a personal email from Dr. Jafar Dhia Jafar, who led the scientific events of Iraq's nuclear program since 1981.

Dr. Jafar mentioned in his email that Iraq submitted complete and comprehensive reports to the Security Council in December 2002 concerning the full details of Iraq Weapons of Mass Destruction (Nuclear, Chemical, Ballistic. and Biological).

That has been an Implementation of the United Nations Security Council Resolution 1441 in November 2002. The resolution gave the Iraqi government a final chance to disclose all details of Weapons of Mass Destruction Programs. The detailed reports of the four weapons programs that were submitted of more than 20000 pages. He provided me with the table content of the complete report on the nuclear programme submitted to the United Nations Security Council and the executive summary report.

Still, the United States of America and Britain started the war on March 19, 2003, without authorization from the UN Security Council. Iraq was occupied under the pretext of the existence of weapons of mass destruction. Strangely, no one has since addressed the validity of the reports submitted by the Iraqi side, which were accurate, comprehensive, and meet the relevant requirements of the resolution.

Appendix A contains a table content of (CAFCD) the Currently Accurate, Full and Complete Declaration (CAFCD) of the Past Iraqi Nuclear Program (INP), consisting of 1409 pages. Copies of which were kept in the Iraqi National Monitoring Directorate.

Appendix B contains the Extended summary of (CAFCD) of 133 pages.

Both (CAFCD)and the executive summary report were published as it was received.

Hence, it's important to be alerted.

APPENDIX A

TABLE OF CONTENTS OF (CAFCD) SUBMITTED TO UNITED NATIONS SECURITY COUNCIL ON 7 DECEMBER 2002

Currently Accurate, Full and Complete Declaration (CAFCD) of the Past Iraqi Nuclear Program (INP)

Contents (Contd.)
PART II a (363 pages)

CHAPTER 4 : TECHNOLOGIES

Contents (Contd.) PART II b (318 pages)
CHAPTER 4 : TECHNOLOGIES

Contents (Contd.)
PART III a (336 pages)
CHAPTER 5 : DEVICE DEVELOPMENT

Contents (Contd.)

Contents (Contd.)

Contents (Contd.)
PART III b (161 pages)
CHAPTER 5 : DEVICE DEVELOPMENT (Contd.)

Contents (Contd.) PART IV (136 pages)
CHAPTER 6 : MISCELLANEOUS ACTIVITIES

Contents (Contd.)

Contents (Contd.) PART IV (contd.)

CHAPTER 7 : SUPPORT FOR NON-NUCLEAR GROUPS & ACTIVITIES

CHAPTER 8 : NUCLEAR MATERIALS

CHAPTER 9 : DOCUMENTATION

CHAPTER 10 : MOVEMENT AND DESTRUCTION OF EQUIPMENT AND MATERIALS

Contents (Contd.)
PART V (119 pages)
CHAPTER 11 : ACHIEVEMENTS

11.2 <u>Centrifuge Enrichment And Gaseous Diffusion</u> 24/119

11.3 <u>Nuclear Device Development</u> 49/119

Contents (Contd.)

Contents (Contd.)

APPENDIX B

(CAFCD) Extended Summmary

**Currently Accurate, Full and Complete Declaration of
the Past Iraqi Nuclear Program Extended Summary**

<u>A brief historical account of the Iraqi Atomic
Energy Commission (IAEC)</u>

The Iraqi Atomic Energy Commission (IAEC) was established in 1956 as a body entrusted with fostering and conducting research, development and training in nuclear science and technology. The IAEC, together with other scientific bodies in Iraq, was established to satisfy the country's basic needs and its aspirations towards scientific and technological progress as a means for economic, social, and cultural development.

During the period 1956 to 1965, the main activity of IAEC focused on building up its basic infrastructure, including training personnel and providing scholarships for students in various fields of nuclear science and technology.

In the early sixties, an economic agreement with the former Soviet Union resulted in constructing the IRT-2000 swimming-pool type light water research reactor, a radioisotope production laboratory, and other supporting facilities such as a small mechanical workshop and a library. These facilities were built at the Tuwaitha site about 20 km southeast of Baghdad. The reactor went critical in early 1968, and the production of radioisotopes commenced late 1969. These facilities later formed the nucleus of the Nuclear Research Canter (NRC). Research work conducted at NRC extended over many fields such as analytical and radiochemistry, nuclear and solid-state physics, health physics, agriculture, biology and engineering.

Iraq signed and ratified the nuclear non-proliferation treaty (NPT) in 1968 and the safeguards agreement in 1972. During the seventies, and per the NPT and the safeguards agreements, Iraq signed many protocols with various countries for the establishment of facilities aimed at the development of nuclear science and technology.

The first agreement was with the former Soviet Union for upgrading the thermal power of the IRT-2000 reactor from 2 MW to 5 MW (IRT-5000). This upgrade was completed in 1978. The supplier (former Soviet Union) changed the reactor fuel from EK-10 and EK-36 of 10% and 36% enrichment to the TBC type of 80% enrichment. This reactor was used until Jan. 1991 to research nuclear physics, solid-state physics, and radioisotopes production. During the device development program, this reactor was also used for the production and isolation of small quantities of ^{210}Po, ^{239}Pu and ^{3}H.

The second agreement, signed on Nov. 18, 1976, was with a consortium of French companies with the approval of the French Atomic Energy Commission (CEA) and the French Government for the construction of the 17th of July project. This project included the Tammuz-1 reactor, a swimming-pool type MTR reactor of 40 MW power providing a high thermal neutron flux of up to $3'10^{14}$ n/cm^{2}s. This reactor was intended to research nuclear physics, solid-state physics, and testing materials (including fuels) normally used to manufacture components for nuclear power plants. It was also designed to be used for the production of radioisotopes. The reactor's fuel was of a sandwich U-Al alloy with an enrichment of 93%. This project also included the construction of the Tammuz-2 reactor, which was a low power swimming-pool type reactor of 500 kW thermal power-producing a neutron flux of $2x10^{12}$ n/cm^{2}s. This reactor was to serve as a neutronic mock-up for the Tammuz-1 reactor to perform measurements on the core, power output, calibration of control rods, measurement of neutron flux distribution, and study the effects of various in-core experiments on reactivity. The reactor pair (Tammuz-1 and 2) had many similarities to the Osiris and Isis reactors operating at CEN Saclay near Paris since 1966. The main differences being that Osiris was of 70 MW thermal power, and Tammuz-1 was equipped with horizontal beam neutron channels and a heavy water tank (of volume 0.7 m^3) with a cold and a hot neutron source for physics experiments. The 17th July project also included materials testing hot laboratory (named LAMA), which was intended for conducting destructive and non-destructive tests on materials after irradiation in the Tammuz-1 reactor. Other parts of this project included a workshop designed to assemble and test the in-core experiments that would be irradiated in the Tammuz-1 reactor. A radioactive waste treatment station (named RWTS) was also included in the project. It was designed to treat low and medium-level radioactive liquid waste resulting from the reactor coolant and other liquid waste from various activities at Tuwaitha.

On Jan. 13, 1976, a third agreement was signed with CNEN, the Italian Nuclear Energy Committee, for scientific and technical cooperation in the peaceful utilization of atomic energy. As a result of this agreement, a contract was signed with Snia Techint, with the approval of CNEN, to construct a research scale radiochemistry laboratory for the development of flow sheets and capable of handling plutonium in gram quantities. This was completed in 1978. IAEC signed in 1978 a contract with Snia Techint, with the approval of CNEN, for the construction of the 30th July project. This project comprised an experimental fuel fabrication laboratory, a technological hall for chemical engineering research, a material science laboratory and a laboratory for producing radioactive isotopes and medical diagnostic kits.

It should also be mentioned that the Iraqi Ministry of Industry and Minerals (MIM) had signed, during 1975-1976, contracts with the Belgian Sebetra company for the utilization of Iraqi phosphate ore at Akashat to produce phosphoric acid leading to the production of various phosphate-based fertilizers at Al-Qaim. The Belgian Mebshem company started in 1982 the construction of a unit for the extraction of uranium from phosphoric acid.

A review of the nuclear programs of other countries in the region, particularly Israel and Iran, shows that the first had obtained from France during the late fifties and early sixties a high thermal power natural uranium graphite reactor (which was upgraded in the early seventies) and an adequate spent fuel reprocessing plant, thereby facilitating weapon's grade plutonium production. Iran had adopted an ambitious plan to install nuclear power plants (NPPs) up to 20000 MWe by the year 2000. Two NPPs supplied by the German firm KWU (Bushir-1 and 2 of 1200 MWe each) were nearing completion, and a further two NPPs of 900 MWe each were being constructed by Framatome of France.

The progress of work of the July 17th project was satisfactory until Israel planned and executed on April 7, 1979, a sabotage operation only a few days before the reactor core block was to be shipped to Iraq. Seven Israeli agents broke into a warehouse belonging to CNIM (one of the contracting companies) in the French port of La Seyne-Sur-Mer, near Toulon. They blew up the core block in an attempt to impede the project implementation. Fortunately, a detailed assessment of the impact of this incident on the progress of work showed that the damaged core block could be repaired and would not significantly affect the completion date for the project. This incident, together with the aggressive campaign in the American and some European news media, perpetrated by Zionist influences that had started shortly after signing this contract and steadily increased in intensity culminating in the Toulon incident, led IAEC to study the feasibility of the utilization of the available reactor and those under construction for the production of plutonium independently. This assessment, conducted during the second half of 1979, was intended to determine if Israeli allegations that this new reactor could produce more than 12 kg of Pu per year had any scientific foundations. The study conducted by IAEC in this respect concluded the following:

(i)   The most efficient utilization of the IRT-5000 reactor would not produce more than 250 g of Pu per year by irradiating natural or depleted uranium within the reactor core.

(ii) The maximum production of Pu by irradiating natural or depleted uranium in the Tammuz-1 reactor core could not be more than 2 kg per year.

Therefore, the Israeli contentions were unfounded since both reactors could not produce more than a couple of kilograms of Pu per year. Moreover, both reactors were subjected to full IAEA safeguards. Furthermore, Iraq did not possess the required spent-fuel reprocessing facilities of an adequate scale.

This assessment indicated unambiguously that the reactors could not in any way be used practically for Pu production, thereby refuting claims by Israel that Iraq intended to use its reactors and other facilities for the production of Pu to produce nuclear weapons.

The aggressive propaganda against the Iraqi nuclear program continued and developed into a terrorist campaign against IAEC personnel and those working for foreign suppliers. Two senior IAEC staff members were murdered in Paris on June 13 and Dec. 18, 1980, and a third in Geneva on June 9, 1981. Also, on Aug. 7, 1980, four bombs were exploded during the same night, two in the Rome office of Snia Techint, one in the Milan home of Snia's director and the fourth in the Paris home of a French reactor consultant injuring him and his wife.

This propaganda and terrorist campaign mounted by Israel reached a climax on June 7, 1981. At dusk, Israel mounted an air raid on Tuwaitha and destroyed the Tammuz-1 reactor when it was nearing completion. This raid took place while Iraq was preoccupied and fully engaged in repelling Iranian aggression.

This was the first time a regime known to possess nuclear weapons covertly, and not a signatory to the NPT, mounted an air raid on a signatory country, thereby destroying safeguarded nuclear installations. Moreover, this Israeli aggression was perpetrated despite Iraq and France have signed an additional agreement stipulated in its article no. 3 the following:

"Both parties guarantee that all facilities, equipment and materials supplied in accordance with this agreement and also all nuclear material produced through the use of the supplied facilities would:

    a.   Neither be used for any military purposes nor in the production of a nuclear-exploding device.

    b.   Neither be transferred to unauthorized persons nor to any party that is not supervised by one of the two parties without taking the written permission of the second party.

    c.   Place the facility under the inspection of the IAEA.

Moreover, Iraq had signed with the IAEA the facility attachment as required by the safeguards agreement for each of the above-mentioned facilities and the design information was provided by Iraq for:

- The IRT-2000 and the IRT-5000 reactors on May 31, 1972, and Dec. 28, 1981, respectively.
- The July-17th project on June 13-June 28, 1980.
- The fuel fabrication laboratory on Feb. 17, 1982.

The Israeli aggression of 1981 clearly proved that neither the NPT and its safeguards nor the bilateral agreements could protect Iraqi nuclear facilities, despite being openly declared, well documented, and built as turnkey projects by international companies. It is worth pointing out that had this aggression not taken place, IAEC would have been fully engaged in its program of research and development utilizing the 17th July and 30th July facilities leading up to the implementation of nuclear power plant (NPP) projects to diversify electrical energy production. Moreover, the French consortium of companies agreed to rebuild the damaged Tammuz-1 reactor. A contract was signed on March 19, 1983, but it remained pending approval of the French government, which never materialized.

Initiation and Objectives of the Past Iraqi Nuclear Program (INP)

2.1 Initiation of the INP

Israel's criminal actions in challenging, frustrating and destroying Iraq's legitimate efforts to acquire nuclear facilities, in accordance with NPT and under IAEA safeguards, created within IAEC a strong conviction that it had to adopt a new policy of pursuing an independent course based on self-reliance. It was considered necessary to conceal this new course to avoid another pre-emptive Israeli aggression. The IAEC had to pursue one of the following two courses:

(i) Developing a reactor program that requires either heavy water or low enriched uranium (LEU) and a spent fuel reprocessing capability of significant Pu production capacity.

(ii) To develop a uranium enrichment technology of significant production capacity of highly enriched uranium (HEU).

It was believed that success in either of these two options would pave the way later to further development on NPPs and other nuclear-related projects in cooperation with the international community. It would also serve as a deterrent against further attacks on Iraqi nuclear facilities. IAEC took the necessary steps to produce enriched uranium and initiated comparative studies of enrichment technologies.

When IAEC concluded that the Tammuz-1 reactor would not be rebuilt, it conducted 1985 a study, project 182, centered around the basic design of a 40 MW_{th} nuclear reactor using natural uranium fuel and heavy water as moderator and coolant. The study extended to other supporting activities that included:

(i) Studies of the flow sheet for fabrication of the required type of fuel.

(ii) Assessment of heavy water production methods. The quantity required for the start-up was estimated to be approximately 40 tons.

(iii) Other related studies are required by the reactor design group.

This study supported an earlier premise that to embark on such a program would require intensive research and development activities as well as a large construction site that could not be executed covertly.

Basic research and development continued at NRC in physics, chemistry, agriculture, biology, and the production of radioactive isotopes and various medical diagnostic kits. Results were published in national and international journals and periodicals, in conferences, and in the annual reports of the IAEC.

Development of Strategies of the (INP)

1.1.1. Initial planning

A comparative study of the various enrichment technologies conducted during the second half of 1981 showed:

Enrichment Technology	Advantages	Disadvantage
1. Electromagnetic Isotope Separation (EMIS).	1. Open and reasonably well documented in the open literature. 2. There is no basic scientific or technical problem that ought to be more or less invented. However, R & D is definitely required to achieve significant production. 3. The computational software and main equipment are not on the trigger list, and therefore can be procured relatively easily. 4. Design and manufacture of the prototype main equipment can be achieved with the available resources and skills. 5. The feed uranium compound is relatively easy to produce and handle. 6. HEU can be achieved with only two stages of enrichment. 7. A batch process where outage of one or more separators does not affect the operation of other separators. 8. Could use Low Enriched Uranium (LEU) as feed with a substantial increase in productivity.	1. Magnets are of relatively large size and mass. 2. Labor intensive 3. Uneconomic for large output. 4. Necessity of recovering feed material from separator internals. 5. Large number of sub-systems.

	9. Many developments in technology occured which could usefully be applied to improve the performance of EMIS, in particular the advances in using on-line computers for process control.	
2. Gaseous Diffusion	1. Commercially proven. 2. Could be implemented with a much reduced number of stages to produce LEU as feed for EMIS 3. Rugged and proven technology.	1. Basic know-how is unavailable. The barrier, for example, has to be produced starting from basic principles, thereby requiring a large effort of R. & D. without a guarantee of success. 2. Requires a large number of rotating machines such as compressors and blowers that are on the trigger list and could not be manufactured locally with the resources available at the time. 3. The uranium feed material is more difficult to produce and handle. 4. Requires a large cascade of 3500 stages to produce HEU. 5. Uneconomical for the production of small amounts of HEU
3. Gaseous Centrifuge : This technology was considered to be even more difficult than gaseous diffusion due to special materials required and the high speed of rotation and their associated mechanical problems.		
4. Laser isotope separation (LIS).	Relatively high stage enrichment factor.	1. Basic know-how is unavailable.

		2. Not a proven production technology in significant quantities. 3. Requires advanced technologies.

A decision was taken at the end of 1981 to embark upon a uranium enrichment program with the following strategy:

1. To establish an enrichment technology leading up to the production of HEU.
2. To adopt EMIS as a primary option and to proceed in its development in three phases:
   (i) To build research units in the first phase.
   (ii) To build production scale units in the second phase.
   (iii) To build production units to achieve 15 kg U/year of 93% enrichment with natural uranium as feed.
3. To adopt gaseous diffusion as a second option with the following objectives:
   (i) To develop a suitable barrier.
   (ii) To develop the specifications of suitable compressors and blowers and to operate them in unit cells.
   (iii) To build a cascade to produce LEU of around 4% enrichment and about 5 tons U/ year to serve as feed for EMIS.
4. All activities were to be based on self-reliance.
5. Activities were organized in such a way to ensure the following criteria:
   (i) A thorough assessment of the published information had to precede any project.
   (ii) Research and development projects were adopted to achieve an adequate understanding of the know-how in the main and supporting technologies.
   (iii) Work had to start on the laboratory scale whenever possible, then be utilized in the pilot scale. The final production scale was to be implemented based on the results of the preceding two stages. These could be conducted in parallel. In other

words, at the time of initiating the first phase of experimental activity, a pilot plant was started. Inherent in this mode of work was utilizing assumptions and parameters that have not been tested adequately or verified experimentally. In many cases, the work on the production plant would be started at almost the same time as the initiation of the experimental work. In such cases, the margin of ambiguity and risk was even greater. As the work progressed, modifications and changes would have to be introduced.

6. Training of technical staff on operation and maintenance of the machines and equipment purchased was included as part of the purchase order whenever necessary.

7. All research and development activities were to be performed at the Tuwaitha site. Production plants or supporting manufacturing plants were to be established in other locations. Such places were to be carefully selected. The main criteria for site selection were:

(i)   To be topographically fortified against aerial attack or sabotage.

(ii) To offer good communication facilities.

The chosen sites would be scattered geographically so that a simultaneous attack would require considerable resources.

Procurement and Production of uranium compounds and Balance of Nuclear Materials

Material declared and subjected to IAEA Safeguards

a. Depleted uranium

In 1979, Iraq imported from Italy 6,005 kg of depleted uranium as UO_2 powder. The material has been verified and fully accounted for and remains in Iraq, under the control of the IAEA, at Location C (a storage facility in Tuwatha) in the same form that was imported.

Natural uranium

In 1979, Iraq imported 4,006 kg of natural uranium from Italy as UO_2 powder and 508 kg uranium as UO_2 in pressed fuel pellets. The UO_2 powder and the pellets were used in the Experimental Research Laboratory for Fuel Fabrication (ERLFF) for research and development activities. Of the 4,514 kg uranium received, 4,323 kg uranium have been accounted for, leaving 191 kg not accounted for. This amount is less than the declared accumulation of "material unaccounted for" and measured discards over the period 1982 to 1990 and maybe considered consistent with the nature of operation of this facility. The balance of this material has been verified and fully accounted for and remains in Iraq, under the control of the IAEA, at Location C.

Low enriched uranium

In 1982 Iraq imported 1,767 kg of uranium from Italy enriched to 2.6% in U-235 in UO_2 powder. The material has been verified and fully accounted for and remains in Iraq, under the control of the IAEA, at Location C, in the same form that was imported.

Highly enriched uranium

Iraq's inventory of research reactor fuel which was imported from either Russia or France, contained almost 50 kg of highly enriched uranium, based on pre-irradiation values. All of Iraq's inventory of research reactor fuel, as listed in Table 4.1, was fully accounted for and removed from Iraq -the last consignment having been shipped in February 1994.

Table 3.1
Iraq's research reactor fuel inventory as verified by the IAEA on 19/20 November 1990

Enrichment% U-235	Number of elements	Irradiation Status	Uranium content (kg)	U-235 content (kg)	Comments
93	1	Fresh	0.417	0.389	Test element
	38	Irradiated	11.874	11.050	Very low burn-up
80	68	Fresh	13.722	10.998	
	62	Irradiated	12.379	9.978	2-12 years cooled
	34	Irradiated	6.812	5.482	Reactor core fuel
36	10	Fresh	3.538	1.272	
	3	Irradiated	1.002	0.360	> 8 years cooled
10	69	Irradiated	87.760	8.776	> 8 years cooled

Mass data are not corrected for burn-up.

Procurement of yellowcake and uranium dioxide

From 1979 through 1982, Iraq procured yellowcake from both Portugal and Niger and uranium dioxide from Brazil. At that time, neither Niger nor Brazil were party to the NPT, nor had either country concluded a comprehensive safeguards agreement, which would have required notification to the Agency of the transfers of such material to

Iraq. Portugal, a party to the NPT but without a comprehensive safeguards agreement, notified the Agency of the transfers to Iraq.

The yellow cake procured from Portugal was supplied in two batches. Batch 1, received on 20 June 1980, consisted of 429 drums containing 138,098 kg of yellow cake and batch two, received as three consignments from 17 May 1982 through 20 June 1982, consisted of 487 drums containing 148,348 kg yellow cake. Iraq notified the IAEA after the material was received by letters dated 6 August 1981, 1 June 1982 and 21 July 1982, which confirmed the complementary notifications made by Portugal at the time of shipment. Iraq's entire holding of the material of this origin was verified against comprehensive packing lists provided to the IAEA by Iraq, detailing the original production lot number together with weight data for each drum. Verification measures involved weighing, non-destructive assay, sampling and analysis from which it was concluded that all of the yellow cake received from Portugal was fully accounted for and remained intact, as shipped, except for the loss of about 40 kg from a drum damaged during Iraq's salvaging activities in 1991. This material remains in Iraq, under the control of the IAEA, at Location C, in the same form as it was received.

The yellow cake procured from Niger was also shipped in two batches. Batch one, received on 8 February 1981, consisted of 432 drums containing 137,435 kg of yellow cake and batch two, received on 18 March 1982, consisted of 426 drums containing 139,409 kg yellowcake. Iraq notified the IAEA after receiving the first consignment in a letter dated Aug. 6, 1981, but did not provide notification after receiving the second consignment. Iraq's entire holding of material of this origin was verified against comprehensive packing lists for both batches, provided to the IAEA by the Iraqi counterpart, detailing the original production lot number together with weight data for each drum. Verification measures involved weighing, non-destructive assay, sampling, and analysis. It was concluded that all of the yellowcake received from Niger was fully accounted for. This material remains in Iraq, under the control of the IAEA, at Location C, in the same form as it was received.

Iraq declared that there had been two receipts of UO_2 from Brazil, the first in August 1981, consisting of 7 914 kg UO_2 in 120 drums, and a second lot received in the first half of 1982 consisting of 128 drums containing from 17,300 to 19,200 kg UO_2.

The IAEA action team (AT) undertook an extensive verification effort involving weighing, non-destructive assay, sampling and analysis, and microscopic examination of the physical form and properties of a comprehensive series of samples of this material. The task was finally completed in July 1994 when, with the co-operation of the Government of Brazil, it was possible to confirm the origin of the UO_2 based on the chemical and physical characteristics determined by the IAEA. It was also possible to gain confirmation of the amount of material shipped to Iraq. These data enabled the IAEA to verify and balance Iraq's declared usage against the material remaining on inventory. Of the 24,260 kg UO_2 received by Iraq from Brazil, 3,600 kg was used to produce UCl_4, UF_4, and uranium metal - the rest has been verified and remains in Iraq, under IAEA control, at Location C.

The Al Qaim uranium recovery facility

The phosphate rock deposits of western Iraq contain uranium in the range of 50 - 80 ppm. A large deposit at Akashat is mined to supply a phosphate fertiliser plant at Al Qaim, some 150 km distant. During the period 1982 to 1984, a plant (Unit 340) for the extraction of uranium from the process, phosphoric acid of the process was constructed and commissioned. Operating at design capacity, the plant should have produced 103 tons of uranium per year -equivalent to 146 tons of yellow cake - assuming 317 operating days and processing 3,600 m^3 per day of phosphoric acid containing 75 ppm uranium at a recovery efficiency of 93%. Over its six years of declared operation by the Al Qaim Phosphate Establishment of MIM, the plant should have produced about 600 tons of uranium contained in nearly 900 tons of yellow cake. However, the plant performance was sub-standard. It produced only 109 tons of uranium in 168 tons of yellow cake, i.e., less than 20% of the plant's design capacity was achieved.

The investigation by the IAEA AT of this apparent inconsistency was greatly facilitated by the presence of a set of operating records – daily production reports - covering the period from 1986 until 1990 and containing day-by-day data on input and output phosphoric acid flows and their respective uranium contents, the relative levels of two key chemical tanks and the number of drums (including drum serial numbers) of yellow cake produced.

The IAEA undertook an extensive evaluation of these data to assess the consistency of the daily operational data with the yellowcake production. Based on sampling at the Akashat mine, a relationship was derived between the uranium and phosphorous pentoxide content of the ore, which enabled the calculation of uranium in the input acid stream. On this basis, it was possible to derive a theoretical estimate of the plant production which was in very good agreement with the declared production.

This analysis also showed that the poor performance of the plant was due to the low assay of the feed acid (60% of design value), the inability of the acid plant to meet the design flowrate of 3,600 m^3 /day (i.e. about 50% of design flow-rate), failure to sustain the 93% design recovery efficiency (actual values were typically 78%) and the fact that the plant on average operated only 214 days per year as opposed to the design value of 317 operation days per year.

The Al Jazira uranium conversion facility

The Al Jazira uranium dioxide and uranium tetrachloride UCl$_4$ production facility, located west of Mosul in northern Iraq, combined a UO$_2$ plant of 185 ton/year design capacity, designated as Project 212 and code-named the "Wax Plant," and a UCl$_4$ plant of 105 ton/year design capacity, assigned Project 244. Both plants sustained considerable damage through aerial bombardment and were thus rendered inoperable in January 1991.

UO2 production

The UO$_2$ production plant was based on designs provided by a Brazilian company. The plant, which was constructed by Iraq in the period July 1985 to July 1989, was based on the well-proven technology involving the dissolution of the input yellowcake in nitric acid followed by multi-stage solvent extraction, ammonium diuranate precipitation, its filtration and calcination to uranium trioxide, from which the UO_2 was produced through hydrogen reduction. The plant began its commissioning phase of operations on 5 July 1989, continuing to January 1990.

This phase was beset with difficulties, and the plant operating records show that only 8,879 kg UO_2 was produced. The plant went into routine operation in February 1990. Apart from being shut down during April of that year, continued to operate until 2 December 1990. All of the available Al Qaim yellow cake had been processed. It was necessary to prepare the plant to process either Niger or Portuguese yellow cake. During December and early January 1991, there was a sporadic operation to clean up waste and scrap and prepare the process for new feed material of different chemical forms.

The Al Jazira UO_2 plant produced 420 drums containing 99,457 kg UO_2 (86,607 kg uranium). Of these 420 drums, five were used for UCl_4 production at Al Jazira, four were used for UCl_4 production in the Chemical Engineering laboratory (Tuwaitha Building 85), and two were used for uranium metal production in the Experimental Research Laboratory for Fuel Fabrication (ERLFF- Tuwaitha Bldg 73). The remaining 409 drums are currently stored under IAEA control at Location C.

Al Qaim yellow cake containing 98,512 kg of uranium was received at Al Jazira. It was converted into UO_2 containing 86,607 kg uranium, resulting in a difference of 11,905 kg uranium. This difference has been investigated in detail. It is estimated that 10,140 kg uranium can be accepted in waste products and damaged plant components, leaving 1,765 kg uranium unaccounted for. This figure is deliberately conservative and could be reduced if greater allowance were to be made for losses resulting from accidental dispersal through Iraq's activities, losses to solvent extraction fluids and losses through dispersal resulting from the aerial bombardment.

UC14 production

The UCl_4 production plant, Project 244, was constructed at the Al Jazira site based on design and operating experience gained from the UCl_4 Pilot plant (Project 242) built and operated in Building 85 at Tuwaitha.

Construction of the Al Jazira plant started in February 1988, and operations commenced on 1 February 1990. The plant consisted of two parallel production lines with a combined capacity of 105 tons/year of UCl_4 and only one line was operational.

The UCl_4 plant operation was limited to 72 hours during February 1990, when it was used to produce 1,200 kg UCl4, containing 780 kg uranium from an input feed of 1,036 kg UO2 containing 901 and generated waste containing 121 kg uranium. Following this brief period of operation, the plant was shut down for maintenance and repairs. It was never again brought back into operation. All of the UCl_4 produced at AI Jazira is stored, under IAEA control, at Location C.

Although it seems inconsistent that the plant would be shut down after only a few days of operation, it should be recalled that the plant had been commissioned well ahead of the need for its contribution to the supply of UCl_4 for the EMIS programme. The commissioning of separators at the Tarmiya EMIS facility began in February 1990. Only eight separators were in partial service before operations were interrupted by the aerial bombardment in January 1991.

Uranium pilot plant development at Tuwaitha

The principal production and use of uranium compounds at Tuwaitha took place in three locations:

- Chemical laboratories (Building 15B) processed Brazilian-origin UO_2 to produce UF_4, uranium metal and UF_6.
- Experimental Research Laboratory for Fuel Fabrication (ERLFF Building 73) processed Brazilian origin UO_2, AI Jazira origin UO_2 and AI Qaim yellowcake to produce UO_2, U_3O_8, UO_3, UO_4, UF_4 and uranium metal.
- Chemical Engineering Research laboratories (Building 85) processed Brazilian origin UO_2 and AI Jazira origin UO_2 to produce UCl_4.

The development of Iraq's capabilities for producing and casting uranium metal originated in Tuwaitha in the middle of 1986. The first phase of this work, which continued through March 1987, was carried out in Building 15 and involved some 30 experiments involving the magnothermic reduction of UF_4. The experiments produced discs of uranium metal of eight-centimetre diameter, having individual weights in the range 600 to 900 grams - 19 such discs remain on inventory at Location C. The experimental work was discontinued in Building 15. Work was not resumed until the beginning of 1988, when facilities in Building 73 were then utilized for the task. The early work in this second phase concentrated on the development of methods to improve the purity of the UF_4 feed material, and it was not until November 1988 that uranium metal Production recommenced. The metal produced in this phase was again in disc form but somewhat thicker - termed "derbies" to distinguish them from the previously produced "discs" - and typically weighed 1.3 kg each.

Phase three involved continued efforts to improve the purity of the UF_4 feed material and a change in the physical form of the produced uranium metal to a solid cylinder of about 5cm diameter and similar length with a typical weight of 1.5 kg.

By late-1989, this research and development established the capability to produce uranium metal of high purity with relatively small process losses. Based on this capability, a larger scale plant was designed and constructed in Building 64 at Tuwaitha to produce 20 kg of uranium metal per day. The plant was still under commissioning in January 1991 when Building 64 was heavily damaged in the bombardment of Tuwaitha. Despite the severe damage to the building, much of the equipment, which was general-purpose in nature, was salvaged and is currently located at the Al Zahf Al Kabir metallurgical facility in the Taji area is subject to ongoing monitoring and verification.

Some 1,150 kg of natural uranium metal was made from 1986 to January 1991, of which 1,000 kg remains in Iraq under IAEA control. About 150 kg was used in metal purification and melting and casting experiments at Tuwaitha and Al Atheer. The most interesting pieces cast were a 5 cm diameter sphere and a small number of 5 cm diameter hemispheres. Except for 10 small uranium bullets and 9 cast rods, all castings and machined uranium pieces were unilaterally destroyed by Iraq's dissolution in HNO_3. Examination of the bullets and bars indicates only rudimentary melting and casting capabilities. As claimed by Iraq and supported by PC-3 program documentation, Iraq expected that considerable improvements would be achieved by utilizing the more advanced equipment soon to be installed at Al Atheer. Much of that equipment was blocked by the export embargo imposed by Security Council resolution 661 (6 August 1990), and all key equipment that was installed at Al Atheer was subsequently destroyed under IAEA supervision.

Exploration of UF_4 and UF_6 production technology spanned the period 1981-1985 and, in 1986, led to the design of Project 206. This project was based on a fluidised bed reactor using anhydrous hydrofluoric acid to produce 2 kg/day of either UF_4 or UF_6. Before construction was completed, Project 206 was modified to produce 1-2 kg UF_4/batch and was renamed Project 231. However, the modified equipment was never operated, and attention was focused on rotary kiln technology.

Project 226, based on rotary kiln technology, was constructed and commenced operation in mid-1986. This project used UO_2 of Brazilian origin as the feed material, which was reacted with Freon 12 as the fluorinating agent, to produce UF_4. Project 226 was operated intermittently until 1991 and produced some 250 kg of UF_4. A small quantity of the UF_4 produced was used in 1987 to make uranium metal. Still, the purpose of Project 226 was to provide a secure supply of UF_4 for eventual conversion to UF_6 to satisfy the needs of the gas centrifuge development program. In the event the material was not required and remains on inventory in Location C.

The lack of success with Project 206 also prompted consideration of the utility of batch processes using boat-type reactors, and small-scale experiments were carried out in 1985-1986 using both Fluorox as the fluorinating agent as well as direct fluoridation using fluorine gas. On the basis of this work, the direct fluorination method was selected for further development, and a larger laboratory-scale boat-type reactor unit, with a capacity of 50g UF_6 per batch was constructed in 1986. This unit operated in Building 15B at Tuwaitha until mid-1987, when it was transferred to Rashdiya. The unit was replicated at Rashdiya, and the two units constituted Project 234.

The amount of UF_6 produced by the unit operating at Tuwaitha was 3-4 kg. Both units operating at Rashdiya were about 4 kg. In 1988 a third unit (Project 235) was constructed at Rashdiya, based on Project 234 designs, and this unit was

used to produce a further 500 grams UF_6. Several other Projects for UF_6 production and purification, including Projects 230, 232, 233, 236, 237, 238 and 238A, were considered but did not proceed beyond the design stage.

The total recorded production of UF_6 is about 8 kg, which was hydrolysed to liquid waste except for 500 grams contained in a standard IS cylinder. The hydrolysed waste and the remaining 500 grams UF_6 are on inventory in Location C.

Projects 234 and 235 provided adequate supplies of UF_6 to Support the development work of the centrifuge program. Project 236, exploiting flame reactor technology, provided sufficient UF_6 to support the pre-production development phase.

Research and development work on UCl_4 production and purification at Tuwaitha is well recorded in IAEC/PC-3 documentation. Initial experiments commenced in 1982 in Buildings 9 and 15. Later, circa 1987, they were transferred to Building 85, the Chemical Engineering Research Laboratories, where activities continued until January 1991. Fifteen laboratory-scale research projects and pilot-scale production and purification projects were implemented during the nine years. Different feed materials, including UO2, UO3, U3O8 and UO4.2H2O, were tried as different reaction techniques such as fluidized bed, static bed (boat type), and rotary reactors with rotary reactors liquid, vapor and gas-phase chlorination.

The extensive experimentation culminated in constructing a pilot-scale production unit, Project 242, in Building 85, which used UO_2 as the feed material and gas-phase chlorination. Project 242, which had a production capacity of 20-40 kg UCl_4 per day, commenced operation in 1988 and continued, on a campaign basis, until the end of 1990. Some 5,000 kg UC14 was produced using Brazilian UO2 and Al Jazira UO2 as feed material during this period. Project 242 was very successful, and the chemical and operating experience so gained was used to design the industrial scale UCl_4 facility at Al Jazira.

Three projects, 241B, 245 and 244, were implemented from 1987 to 1990 to establish the capability to meet the purity requirements for EMIS feed material. These projects, all based on sublimation, were used to purify some 1,100 kg of UCl_4.

The nuclear material balance for these Tuwaitha locations (Table 4.1) shows a total receipt of 14,789 kg uranium, of which 13,135 kg uranium has been verified and remains on inventory at Location C. The resultant inventory difference or "material unaccounted for" (MUF) is 1,654 kg uranium, representing 11.2 % of the total receipts. Some components of this MUF comprise strata that are physically present but difficult to verify, with any certainty, such as the Building 73 waste, plant hold-up, uranium losses to metal slag and others for which detailed explanations were provided to IAEA backed up by documentation, such as the hydrolysis of UF_6 and the dissolution of uranium metal. Conservative assessment of these components reduces the MUF to 1,068 kg uranium or 7.2 % of the receipts. Considering that some large inventory strata are inhomogeneous and thus potentially subject to large sampling errors, and given the loss of some material, this MUF value is due to the bombardment and subsequent salvage activities is considered to be entirely reasonable.

Table 3.1.1 Material balance - Tuwaitha uranium projects

Receipts into Tuwaitha uranium projects

Material origi Compound type	Compound (kg)	Uranium (kg)
Brazilian UO_2	3,600	3,150
Al Jesira UO_2	2,504	2,180
Al Qaim yellow cake	14,072	9,459
Total		14,789

Verified accumulated inventory

UO_2		2,186
UO_3		3,188
UO_4		3,667
UCl_4		1, 917
uranium metal UF_4		1,023 226
ADU		598
Miscellaneous		330
Total		13,117

Material unaccounted for (MUF) 1,654 Unverified components of MUF

Hydrolysed UF_6		7
Waste - Building 73		206
Dissolved uranium metal		150
Uranium metal slag		60
Plant hold-up		163
Total		586

Adjusted material unaccounted for 1,086 Iraq's former holdings of research reactor fuel are listed in Table 3.1.

3.2 The production and separation of plutonium

3.2.1 The indigenous reactor - Project 182

a. Background

Project 182 was established in late 1984 to design and construct a natural uranium fuelled, heavy water moderated and cooled reactor of some 40 MW$_{th}$ capacity modeled on the Canadian NRX research reactor. This project was established after rebuilding the Tamuz-1 reactor destroyed in the Israeli air attack of 7 June 1981, which was shelved by the French Companies. Project 182 was intended for the production of plutonium.

b. Development

The design of the reactor did not progress beyond preliminary design studies. No decision had been made as to whether the fuel would be in the form of ceramic oxide or metallic uranium. The priority allocation of resources to the EMIS program had, for all practical purposes, put Project 182 "on hold" indefinitely.

Consideration was given to the conversion of Project 182 to an "open project" and seeking the co-operation of the IAEA or other international bodies to facilitate its implementation. However, a subset of Project 182 dealt with heavy water production. A PC3 report, issued on 22 October 1990, reviewed the two most widely utilised production processes.

The use of the IRT 5000 reactor

The use of the IRT-5000 reactor in reprocessing research and development activities was twofold. Firstly an irradiated IRT-5000 reactor fuel element (10% enriched uranium - EK10) exempted from IAEA safeguards at Iraq's request was reprocessed. Secondly, three fabricated natural uranium fuel elements were irradiated in IRT-5000 and also reprocessed. While it is clear that the IRT-5000 reactor made a useful contribution to the research and development program. It was of limited usefulness as a plutonium production reactor.

The separation of plutonium

A laboratory-scale process line, Project 22, based on PUREX technology, was built and successfully commissioned in the hot cells of the radiochemical laboratory at Tuwaitha (Building 9). Three reprocessing campaigns were carried out from April 1988 through April 1990, the first two of which involved the reprocessing of EK10 fuel pins and the last the reprocessing of pins from three "home-made" (EK07) fuel cassettes. Some five grams of plutonium was separated, and about 11 kg of uranium was recovered through these reprocessing campaigns.

Through Project 22 also, a completed laboratory-scale experiment to produce milligram quantities of plutonium metal employing classical "bomb-reduction" techniques was conducted.

Development of Uranium Enrichment Technologies

As stated in section 2, Iraq's strategy for acquiring weapons-usable nuclear material, established at the end of 1981, was to use electro-magnetic isotope separation (EMIS) as the primary technology.

The initial strategy foresaw developing an industrial-scale plant with 15 kg/year production capacity of highly enriched uranium (HEU - 93%), based initially on natural uranium feed. Gaseous diffusion was chosen as a subsidiary technology with the declared objective of building a plant to produce 5 tons/year of low enriched uranium (LEU) containing 4% U-235 to be used eventually as the feed material for the EMIS plant. Assuming that the EMIS plant could have been optimised to use LEU feed material, combining the two technologies could have tripled the EMIS plant's capacity.

Other technologies such as gas centrifuge enrichment and laser isotopic separation (LIS) were not included in the initial strategy because of their greater technical complexity and dependency on equipment subject to export controls. Nonetheless, LIS and chemical and ion-exchange uranium enrichment processes were explored. However, only centrifuge technology was taken beyond laboratory-scale exploitation.

In 1987, the further development of gaseous diffusion technology was assigned reduced priority. The released resources were assigned to the development of gas centrifuge enrichment.

- **Electro-magnetic isotope separation (EMIS)**

The EMIS development program was organised into three phases. The first phase concentrated on research and development activities using "R40" isotope separation magnets. Three units were designed to have ion-beam paths of radius 40 centimetres, were 1:3 scale versions of the anticipated production scale units. Phase one was established in Tuwaitha and continued over the period 1982 through 1987. It involved constructing and operating an electromagnet (Project 101) and two different magnet/separator systems (Projects 102 and 103), all of which were in operation in Building 85 from the beginning of 1985.

The second phase, which overlapped phase one, commenced in 1983 and reached an experimental stage in 1987. Phase two was devoted to developing R50 and R100 pre-production-scale units (Project 104), as well as 1:5 magnetic field-scale model (Project 105), which were used to investigate multi-magnet series operation as an analytical tool for the production phase configuration. Starting from 1985, one R50 and three R100 magnet/separator systems were built and installed in Building 80 at Tuwaitha and were operated until 1991. moreover, phase two included the design and implementation of project 106, which was conceived to study the behaviour and operational aspects associated with multiple arc ion sources. According to program progress reports, none of these separators achieved more than 20% of their design capacity. This performance is in keeping with the fact that the total production of enriched uranium from the development separators at Tuwaitha was only 640 grams, with an average enrichment of 7.2%.

The design work for the third phase, the production phase, which proceeded concurrently with the other two phases, was finalised in 1987 and foresaw two identically equipped industrial-scale plants, Al Tarmiya and Al Sharqat, each with 70 R120 separators for the production of uranium enriched to about 18% and with 20 R60 separators for the production of HEU (93%). The design production of each facility was 15 kg HEU per year, based on natural uranium feed, with the potential of more than a three-fold increase in that production by using LEU as the feed material.

A Yugoslav civil engineering contractor was employed to construct many of the principal buildings at Al Tarmiya. Still, Al Sharqat was being built by an Iraqi company.

The installation and commissioning of R120 separators at Al Tarmiya commenced at the beginning of 1990 and that, by Jan. 1991, a total of eight R120 separators were in preliminary operation. Preparations had begun for the second group of seventeen R120 separators to be installed but were not accomplished.

The total enriched uranium produced at Al Tarmiya is about 685 grams at an average enrichment of 3%, equivalent to only about 20% of design, both in terms of mass and enrichment, but is not inconsistent with the reduced performance that might be expected during commissioning.

The operation was interrupted on the December 15, 1990 and the plant was damaged by the bombardment in Jan. 1991.

Construction of the sister facility at Al Sharqat was about 80% complete at the end of 1990. No EMIS process equipment was installed at Al Sharqat.

- **Gaseous diffusion uranium enrichment.**

a. Background

Exploratory work on gaseous diffusion technology commenced in 1982 with the intention of developing the capability to produce low enriched uranium for use as feed material for the EMIS process. The work had initially concentrated on the development of suitable porous barrier material, on obtaining a theoretical understanding of molecular flow through porous media and on diffusion plant cascade design. By 1985 some progress had been achieved in producing barrier material, thereafter effort was also placed on compressor, diffuser and heat exchanger design. It rapidly became apparent that a very large industrial infrastructure would be required to manufacture these items and that this infrastructure was beyond the national capabilities at that time.

A decision was made in 1987 to revise the mission of the team assigned to this task (Group One) such that priority was given to the exploitation of gas centrifuge technology for uranium enrichment. Some work on the gaseous diffusion the process did continue, although it was limited to research and development on the barrier material and on carrying out practical tests on some compressors that had been procured.

Research and development

Work commenced in 1982 with literature surveys of data on separation barriers, followed by experiments on porous tube manufacture and the characterisation of porous materials. A number of materials, in various forms and deposited by various methods, were investigated during the following three years with little success due to excessive pore size and unsatisfactory flow characteristics. A suitable barrier material was developed in 1988, which overcame these adverse properties, but the barrier tube was still mechanically weak in industrial-scale handling.

In parallel with the above, a survey of compressors judged to be suitable for transporting the process gas was made and specifications were obtained from potential suppliers. Procurement action was taken to purchase compressors from companies in the USA, Germany, France, and the UK and attempts were made to manufacture a compressor casing locally, but these were not successful. In 1987 design drawings of a screw compressor were made by reverse-engineering a screw compressor procured from the UK. However, it was soon realised that the reproduction of its components was beyond the capacity of the existing national engineering resources. Although some attempts were made to secure foreign assistance, nothing materialised. Concurrent with these activities, a facility for testing compressors was built at Rashdiya but was never commissioned due to the change in emphasis of the program in favour of the centrifuge enrichment process.

Theoretical work on diffusion cascade behaviour and calculation of the performance of a total cascade made up of different sized stages acting in "square."

Cascade array was carried out. These calculations were for various cascade sizes ranging from 16 stages in series to 72 stages in series. Theoretical calculations aimed at optimising the geometry and flow parameters of the diffuser were also made.

Facilities were constructed initially at Tuwaitha, then later at Rashdiya, to test the theoretical models of the barrier design and the diffuser. These test facilities included capabilities to check barrier porosity, permeability, robustness and gas flow dynamics for tests with inert gas and with hydrogen fluoride (HF), fluorine (F_2) and the process gas (UF_6). Although a number of facilities to test barrier performance in UF_6 were planned, none were carried out.

Barrier manufacturing facilities were commissioned to investigate the various proposed manufacturing processes, culminating in a laboratory-scale production facility capable of making 18 test barrier tubes per day - several hundred were produced during its operating lifetime. In 1986 plans were made to test a single barrier tube with UF_6. The tests were carried out at Rashdiya in 1988, within Project 365, where one barrier was exposed to UF_6 for about four months, and promising results were obtained.

It was planned to measure the separation factor of a complete single-stage unit, using a mixture of two freons with very different molecular weights. A separation-test facility was constructed at Tuwaitha, but severe difficulties were experienced in assembly due to the lack of robustness of the barrier tubes. Many were broken before an engineering solution was achieved. However, before the facility was commissioned, the entire project was moved to Rashdiya. The facility was dismantled and transferred to the new site but was not rebuilt.

In 1988 a barrier tube suitable for operation in UF_6 was successfully manufactured. The separation performance of a single unit (or stage) was theoretically determined, and planning commenced on Project 366 through which to assess the barrier efficiency of 24 stages operating in series. This plan was not completed, and the project was cancelled in 1989. Two further facilities to measure the separation factor in UF_6 gas of a single diffuser stage unit and 48 diffusers acting in series were also planned. The design of the former was completed but, due to the revised program priorities established in 1987, was not constructed. The design of the latter was not completed, and the project was abandoned at the basic design stage.

- **Gas centrifuge uranium enrichment**

Background

The team responsible for the development of gaseous diffusion technology (Group One) became independent from PC-3 in August 1987. It was renamed the Engineering Design Directorate – eventually to become the Engineering Design Centre (EDC). At the same time, it relocated from Tuwaitha to premises (Rashdiya) in the north-western outskirts of Baghdad, which had formerly been a Ministry of Irrigation research and development establishment.

It was decided to focus the resources of EDC on the development of gas centrifuge enrichment technology to establish a production capacity of 10 kg of highly enriched uranium (93% - HEU) per year by 1994. The facilities on the new site were rapidly expanded, and modifications to existing buildings and new building construction continued until early 1991, as work on the centrifuge enrichment process gathered momentum.

A large number of technical drawings was handed over to the IAEA, which showed the progress of the design of the various types of centrifuge machines considered.

Research and Development

Work commenced in August 1987 with an attempt to develop the oil-bearing (Beams type) gas centrifuge for which extensive design information was available in open US literature. EDC's technical capabilities developed rapidly and, by late 1987, the first oil centrifuge (GS-1) was built and subjected to laboratory trials. Rotational speeds greater than 30,000 rpm could not be achieved due to vibration, high power consumption and vacuum difficulties.

In the face of these difficulties in the Summer of 1988, EDC sought foreign assistance through H&H, a German company already involved in supplying specialist machine tools to MIC.

H&H introduced two foreign nationals who had previously been employed by MAN. This German company had, in the 1970s and early 1980s, been involved in the design, development and supply of centrifuges to URENCO, the European centrifuge enrichment company, which produces low enriched uranium (LEU) fuel for nuclear power plants. During the following two years, the difficulties with vibration due to imbalance and vacuum were gradually overcome as rotor dynamics and bearing know-how was learnt, with guidance from the ex-MAN employees, and by the import of high quality balancing machines and drive units. By mid-1989, a speed of 50,000 rpm was achieved in a vacuum. These mechanical trials were followed by separation tests using a mixture of freon and carbon dioxide to simulate uranium hexafluoride (UF_6) gas, the medium used in the centrifuge enrichment process. The separation tests carried out at a maximum rotational speed of 25,000 rpm gave a separation factor of only 1.04, which was much lower than the theoretical value of 1.09.

By this time, resources assigned to the development of the oil-bearing centrifuge were already being reduced in favour of developing the more efficient magnetic bearing centrifuge, which was commercially exploited on an industrial scale.

The shift of focus from the oil-bearing centrifuge was due to the provision, in the second half of 1988, by one of the ex-MAN employees, of a number of design drawings relating to early development designs of a magnetic bearing (Zippe type) centrifuge. Consequently, EDC applied most of its resources to the design and development of a magnetic-bearing centrifuge based on a maraging steel rotor rotating at sub-critical speeds.

During 1989 H&H introduced a further ex-MAN employee who, in co-operation with one of the original individuals, provided to EDC many detailed design drawings and some 170 technical reports and specifications relating to the production and operation of centrifuges under development by URENCO in the 1970s. This information covered both sub-critical and supercritical centrifuge designs. Also, it included some drawings for a three-meter long supercritical machine under development, by MAN, in the early 1980s.

From late-1988 through mid-1990, EDC produced a series of designs, each one initiated by information or advice deriving from the ex-MAN employees, and proceeded to attempt to manufacture trial quantities of centrifuge components. It was quickly concluded that Iraq's existing manufacturing capabilities were unable to produce the rotating components of centrifuge machines to the required accuracy and quality, and, in the first instance, production was limited to stationary components. A decision was taken to strengthen the industrial infrastructure by importing high-quality, dedicated CNC machine tools, in most instances linking the purchase to the supply of quantities of demonstration components that were to be used for the assembly of development centrifuges.

Machine tool suppliers were approached in Germany, Yugoslavia, and Switzerland. Some orders for small quantities of components were placed with a German company and a UK company, which were not linked to the supply of machine tools. EDC's procurement strategy did not always proceed smoothly as demonstrated by the German customs authorities at Frankfurt airport's impounding of machined maraging steel forgings, finished maraging steel components and CNC machine tools being supplied by a Swiss machine tool company.

In mid-1989, an offer was received and accepted from one of the ex-MAN employees to provide design details of a sub-critical centrifuge based on a carbon fibre composite rotor and supply some trial rotors. Carbon fibre composite had many technical advantages over maraging steel. It had been adopted as the standard material in European commercial gas centrifuge enrichment plants. By the end of 1989, EDC had developed a series of sub-critical centrifuge designs based on the carbon fibre rotor. By early 1990, sufficient components had been procured to support prototype centrifuge production and testing. The procured components included about 50 carbon fibre rotors supplied by the company owned by the same ex-MAN employee who had provided the offer.

In the spring of 1990, the first magnetic centrifuge using a carbon fibre composite rotor was successfully assembled and tested at an operating speed of 60,000 rpm over several months in a mechanical test stand. In mid-1990, this centrifuge rotor was installed in a process test stand, and about 100 hours of operation in UF_6 was achieved during the following 6 months. Although not fully optimised, a separative work output of 1.9 kg SWU/year was achieved with the prototype. A cascade of 1,000 such centrifuges operating continuously for one year would produce about 10 kg of 93% HEU.

No enriched uranium was accumulated during these separation tests since the operation was in total reflux mode, i.e. the enriched material produced was re- mixed with the tail depleted material for re-feeding into the test centrifuge input. The mechanical and process test stands were the only two test stands that were operated. A third test stand designed to accommodate two centrifuges operating in series or parallel, planned for late 1990, was not implemented.

The exploitation of the designs of supercritical centrifuges was limited. It had been done on a spare-time basis, as the bulk of EDC resources were dedicated to the further development of its prototype sub-critical machine and preparations for its large-scale production. The studies on supercritical centrifuge machines were focused on the design of a three-metre machine, simply because the information it had obtained for this particular centrifge design was far more complete than the information it had on a two-cylinder maraging steel rotor design.

Preparation for production

In mid-1989, EDC contracted with both local and foreign companies for the construction of the Al Furat facility, which was to accommodate the factory for the mass production of centrifuges and a pilot-scale cascade hall. It had also planned to build a second large-scale centrifuge facility in the Taji area, which was intended to accommodate a cascade of up to 1,000 centrifuge machines and accommodate a commercial scale UF_6 production plant.

In the summer of 1990, EDC received from H&H a flow-forming machine installed at Al Furat and enabled the flow forming of maraging steel rotor cylinders to commence on a trial basis. Around this time, ancillary equipment for welding and heat-treating maraging steel was also imported. Only a few heat treatment tests were carried out.

Existing buildings on the site were modified, one of which (B03) was used temporarily from Autumn 1990 for production development trials, a further building (B00) was almost complete in its refurbishment and was ready to accommodate CNC machine tools, delivery of which had already begun, when activities were suspended in Jan. 1991. Two large, purpose-designed buildings were under construction and at an advanced stage, although some 6 months behind schedule. One of these (B02) was being constructed by a UK company and the other (B01) by a German company, and both involved cleanroom technology.

Building B02 was to be used for flow forming, component cleaning, quality control and sub-assembly. Building B01 was intended for final assembly, single machine spin testing, cascade pipework manufacture and a demonstration cascade of 120 machines capable of producing about 1 kg HEU/year. To support the construction phase, H&H persuaded a small number of companies that had previous experience in centrifuge manufacture and plant construction as URENCO contractors to run training courses for some EDC staff on corrosion of special steels, pipework fabrication and welding technology.

In parallel with these activities, EDC was actively pursuing carbon fibre composite technology and in 1989 had ordered through the company ROSCH a purpose-built carbon fibre winding machine and a supply of carbon fibre filament and epoxy resin to establish a capability to manufacture carbon fibre composite cylinders for centrifuge rotors. The delivery to Iraq of these materials and equipment was initially prevented by the 1990 prohibition of exports to Iraq, but the delivery of the equipment and materials was made to Jordan in 1991. This had been accomplished through a system of transshipment through an import/export agency in Singapore – the equipment and materials were not imported to Iraq and were handed over to the official custody of the IAEA in Jordan.

Due to delays in the completion of Al Furat, a decision was made to construct an additional building at Rashdiya, including a centrifuge hall to accommodate the pre-production-scale 120 centrifuge cascade. Additional work was also undertaken to adapt part of an existing building at Rashdiya to accommodate a 50 centrifuge cascade as part of the "accelerated program."

- **Chemical and Ion Exchange Uranium Enrichment**

Background

Research and development into uranium enrichment through solvent extraction and ion exchange processes commenced in 1988. The decision to explore these enrichment technologies followed a review by the Iraqi Atomic Energy Commission (IAEC) of known enrichment methods and a similar review of the feasibility of a plutonium production reactor. The relocation and reassignment of Group One, which had been pursuing gaseous diffusion technology within IAEC Department 3000, during the summer of 1987, provided the impetus for these initiatives.

The objective of the investigation of these two additional enrichment methods was to provide an alternative supply of low enriched uranium (LEU) as the feed for the EMIS facilities.

Group Two Activities 2CC and 2CE contributed to the exploration of solvent extraction and ion exchange enrichment.

Chemical Enrichment (Solvent Extraction)

Chemical enrichment by solvent extraction was modelled on the French CHEMEX solvent extraction process, which was well described in the open literature. Only elementary practical work on the CHEMEX process was carried out, but it was sufficient to establish important fundamental factors.

The goal of the chemical enrichment process was to provide LEU feed material (1.5-2.0 % U-235) for the EMIS process. The production scale design called for an annual production of 4-5 tons of LEU (3-4 % U-235) based on theoretical estimates. The production-scale design foresaw about 50 stages and anticipated a separation factor of 1.0025.

Detailed laboratory work was carried out in Tuwaitha pursuing basic studies designed to measure the separation factor, using 30-35% TBP (tri-butyl phosphate) as the extractant in a kerosene diluent, but by Jan. 1991, work had not progressed beyond the laboratory scale.

The strategy adopted was to address practical problems as they arose in scaling up to production scale. It was clear that many significant technical challenges would have to be met.

Imports to support this research were limited to laboratory equipment such as mixer-settlers, pumps, distillation units, and pulse columns. Much of this equipment was destroyed during the aerial bombardment of Tuwaitha. Orders were placed for key pilot plant equipment such as glass columns and mixer-settlers, but the 1990 embargo on exports to Iraq prevented their delivery.

Ion exchange enrichment

Ion exchange enrichment was modelled on the Japanese ASAHI technique, which was well described in the open literature. The objective of this program was to establish the capacity to produce 5 tons of LEU (3% U-235) per year for use as feed material for the EMIS process.

Less progress was achieved in the ion exchange enrichment process than in the CHEMEX process. The laboratory-scale work stopped Jan. 1991.

A total of about 100 kilograms of polyvinyl, phenyl pyridine-based, macro reticular (highly porous) anion exchange resin was produced in 20-kilogram batches over two years.

Experiments carried out using a four-meter long, two-centimeter diameter column achieved a separation factor of 1.0007. The experiments were conducted at a nominal pressure of 4 bar and a nominal temperature of 80 degrees Celsius.

Consideration was given to a combined solvent extraction/ion exchange enrichment process. The output of the solvent extraction process would have fed the ion exchange process with 1.5-2.0% LEU. The output of the combined process would have been 8% LEU, which was again intended to be used as feed material for the EMIS process.

• Laser isotopic separation

The Laser Section (6240) within the Physics Department (6200) of the Iraqi Atomic Energy Commission had in 1981 been directed to work on Laser Isotopic Separation and to study both atomic (AVLIS) and molecular (MLIS) technologies. The activity achieved little more than scrape the surface of either technology. This lack of achievement was due in part to the complexity of the technology and also to the difficulties experienced in obtaining critical controlled equipment, notably copper vapour lasers.

The program neither reached the point of an integrated experiment that achieved any isotopic separation of either elemental uranium or UF_6

nor approached the most rudimentary capabilities in either AVLIS or MLIS technologies.

Two attempts were made to construct a suitable vacuum chamber to facilitate AVLIS experiments. The second of these attempts was successful. The chamber was equipped with an electron beam gun for the vaporisation of uranium metal. One experiment using two-photon excitation was carried out in 1986. Still, it did not produce conclusive results due, it was thought, to lack of precision in the design of the ion optics. A second experiment was carried out in 1989 after having optimised the equipment's internal arrangements on the basis of results obtained from experiments with aluminium metal. The experiment with uranium metal proved to be inconclusive. Further work was abandoned due to the electron beam gun's failure and the low priority assigned to the research program that would not support the procurement of a replacement.

Device Development

Background

No practical steps towards establishing device development commenced until the end of 1987. In early 1987, the Al Hussein Project (HP) was established under the direct supervision of the Minister for Ministry of Industry and Military Industrialisation (MIMI) and comprised a small group of individuals tasked to assess the resources, investment requirements and the time frame to achieve the first nuclear device. The HP delivered a summary report in November 1987, which was met with strong criticism due to its exaggerated and inflated requirements. This led to the abandonment of HP and the establishment, within IAEC, in April 1988, of a device development team known as Group Four.

With the transfer from IAEC, in November 1988, of Department 3000 and its establishment, in January 1989, as PC3, within the MIMI, device development activities were divided between PC3 and Dhafir Project at Al QaQaa. PC3 was responsible for device design, fabrication, and Dhafer Project was responsible for developing and producing detonators and high explosive lenses. The initial activities of Group Four of PC3 were carried out at the Tuwaitha Nuclear Research Centre until May

1990. With the exception of the Theoretical Studies, Reprocessing, and Uranium Conversion Departments, which remained at Tuwaitha, Group Four moved to its new site at Al Atheer.

• **Facilities**

As the principal nuclear research centre in Iraq, Tuwaitha had the facilities and infrastructure for all Group Four activities except for the high explosives' fabrication, handling, and testing. Theoretical studies, based on the use of mainframe and personal computers, electrical design studies, and the development of dedicated instrumentation were carried out in regular buildings at Tuwaitha. Radiochemistry experiments, including the separation of a few grams of plutonium, took place in the hot cells of Building 9. Studies of uranium metal production and casting were conducted as part of the activities related to fuel fabrication and used facilities in Buildings 15 and 73.

Al Atheer was specifically designed to accommodate all technical activities related to nuclear device development, including experiments with high explosives for which an elaborate complex was designed and constructed. The complex included a heavy-duty bunker (Site 100) and an internal explosion chamber (Site 6600). Site 100, which was capable of handling experiments involving several hundreds of kilograms of high explosives, was completed as early as 1989. The design of the internal explosion chamber included a high integrity containment system to prevent the release of radio-toxic materials used in neutron initiators. The construction of Site 6600 was still incomplete when hostilities began in Jan., 1991.

Uranium metallurgy studies and fabrication, both for natural and highly enriched uranium, were to be accommodated in building (6830) equipped with a sophisticated air handling system. Another building (430) was designed to accommodate equipment and facilities for the machining of uranium metal. Both buildings were still under construction at the end of 1990.

A powder metallurgy building, which was already equipped with large industrial hot and cold isostatic presses, was close to completion at the end of 1990. Other buildings were designed for material characterisation, dynamic testing of materials, neutron source testing, device assembly and storage. Dedicated facilities were also provided for civil engineering support activities and mechanical and electrical design activities.

When completed, Al Atheer would have been equipped to develop, fabricate and cold test the nuclear device and its individual components. All the technically significant buildings and the related equipment at Al Atheer were destroyed under IAEA supervision in April and June 1992.

Al QaQaa, which was the main facility for the production of conventional high explosives, detonators and missile propellants, had the infrastructure to support the initial activities of the Dhafer Project in the development of the high explosive package for the device. Al QaQaa held large stocks of imported HMX and RDX and had its own operating RDX production plant.

However, as the work of the Dhafer Project progressed, contracts were signed with foreign suppliers to build turnkey research and development facilities for pyrotechnics and the production of shaped high explosives and associated experimentation. A contract for the construction of RDX and HMX production facilities at a location near Falluja was also concluded.

Civil engineering work began on all these contracts, and some equipment was provided. Still, the August 1990 embargo, imposed by UNSCR 661 (1990), halted all projects before completion.

Existing facilities, including a number of buildings previously used for composite missile propellants, were used for the production of various detonator types and for pressing and casting shaped high explosives.

A location in southwestern Iraq was proposed for underground nuclear testing on the basis of certain criteria. This site was to have been available by the end of 1991, but final selection had not been made, and construction was not started before the program was abandoned.

Research and Development

Group Four's theoretical activities concentrated on studies of the requirements of an implosion weapon "fuelled" by HEU - the study of a gun-type weapon having been abandoned in 1988 because that design was known to require a much larger amount of highly enriched uranium

(HEU) than the implosion design. Group Four nuclear device design relied heavily on information available in the open literature.

Theoretical studies led to the development of various computer codes to evaluate the performance of a given design. These codes were also obtained from open literature and were adapted to the available mainframe and personal computers. Group Four undertook to adapt the codes and develop the physical constants, such as equations of state, neutron cross-sections, and the constitutive models, which assessed the nuclear device development program needed. Although the primary focus was a basic implosion fission design, fuelled by HEU, Group four was aware of more advanced weapon design concepts, including thermonuclear weapons. Group Four also invested significant efforts in understanding the various options for neutron initiators.

In the area of electronic and electrical design, Group four developed its own instrumentation to be combined with imported equipment such as streak cameras and oscilloscopes. Fast electronic components, flash X-ray devices, and sensors of various types were also under development. However, imported equipment was procured when possible. Group four was developing an arming, fusing, and firing system for a 32-point detonation system.

The Dhafer Project followed a largely empirical development program in its work to produce high explosive lenses for the implosion package. Until the first half of 1990, the project concentrated on the use of pressing to form the lenses, but the size limitation imposed by the available equipment resulted in a transfer of effort to high explosive casting technology. Development of plastic bonded explosives did not progress beyond laboratory-scale production.

Single pressed lenses were tested, but no cast lenses had been produced by January 1991, and thus, none had been tested. Four-pi tests or any test of multiple lens arrays were not conducted.

Significant progress was made in developing capabilities for the production, casting and machining of uranium metal. However, Group Four had not progressed beyond casting centimetre-sized test pieces to cast full-scale pieces due to the delayed importation of adequate furnaces.

Nonetheless, success was achieved in casting a uranium sphere of about five-centimetre diameter, several hemispheres of similar size and a small number of rods, weighing 1.2 kilograms per piece, from which to machine "sub-calibre munitions."

- **Missile delivery system**

Consideration of a missile delivery system for a nuclear device commenced as early as 1988 in a meeting attended by a senior deputy minister of MIMI. However, no further interaction took place until the end of 1990, when the need arose to liaise regarding the integration of the nuclear device, which was to have been produced through the "accelerated program" with a missile delivery system.

In Dec. 1990, discussions of the device conceived package that it was too heavy to be delivered by existing Iraqi missiles. Group Four was tasked to modify the design "with a view to reducing the total weight of the projectile to about one ton or less." The long-term plan was for a delivery vehicle based on the engine that was being developed for the second stage of the Al Abid satellite launch vehicle.

The options considered for the "accelerated program" involved either the urgent production of a derivative of the Al Hussein/Al-Abbas missile, designed to deliver a one-ton warhead to a maximum range of 650 km, or to accept the fallback option of using an unmodified Al Hussein missile and to accept a range limitation of 300 km.

The "Accelerated Program"

Background

Hussein Kamel (HK) gave an order on Aug. 18, 1990 to divert from IAEA safeguards the highly enriched uranium (HEU) contained in the fuel of the two research reactors on the Tuwaitha campus of the (IAEC) and to use this material to produce the core of a nuclear device.

This plan referred to as the "Accelerated program," involved the recovery of the HEU from the research reactor fuel and its subsequent conversion to metallic form as raw material for the production of the core of a nuclear device.

Although, the plant for the recovery of the HEU had been built and fully commissioned, the campaign for actual extraction of the HEU from the reactor fuel was not initiated. This fact has been corroborated by the IAEA AT that accounted fully for the entire inventory of HEU reactor fuel.

The inventory of enriched uranium research reactor fuel under IAEA safeguards, as of April 1991, is shown in Table 3.1.

The recovery of the highly enriched uranium – Project 601/603

Project 601 was established in August 1990 with the objective of extracting the highly enriched uranium (HEU) from research reactor fuel for use as the core material of a nuclear device. A chemical plant, based on solvent extraction technology, was designed, and its components fabricated and installed in the hot cells of the Active Metallurgy Testing Laboratory (LAMA), Building 22, in the Tuwaitha site.

The team working on this project had already accumulated experience from its laboratory-scale work on the separation of plutonium from irradiated natural uranium fuel rods. It was confident that it would be able to achieve its objective. The throughput of the plant was designed to accommodate the processing of one, possibly two, fuel elements per day such that the recovery of the HEU from the 69 fresh and 38 lightly irradiated fuel elements could have been accomplished within 2 to 3 months, thus making available some 26 kg of HEU, in the form of UNH containing 22.4 kilograms of the isotope U-235, less process losses.

The next phase of the plan would have involved the processing of the highly irradiated HEU reactor fuel, making available a further 14 kg of HEU containing some 10 kg of the isotope U-235. This project phase would present a greater technical challenge because of the need to remove considerable fission product contamination from the separated uranium - the process losses would most likely be significantly higher.

Calculated data were made from which to estimate the fission product content of 62 irradiated fuel elements (80% enriched) based on tabulated data of the burn-up and cooling-time of each element. These 62 elements, together with the 34 elements remaining in the core of the IRT-5000 reactor, represented the total inventory of 96 irradiated fuel elements of 80% enrichment, as verified by the IAEA on 19 November 1990. Typical fission product content of the much more lightly burnt 93% enriched fuel from the Tammuz 1 reactor was also calculated.

Other less significant phases of the project would have involved the recovery of the uranium from reactor fuel of lower enrichment, much of which was subjected to high burn-up.

The chemical plant's design, fabrication, and installation was completed within a period of little more than three months, which enabled the plant to be commissioned using unirradiated natural uranium solutions during December 1990. The unit was ready to receive HEU feed material in early January 1991, and clearance had been sought from HK to commence actual operations. No such clearance was received, and the fuel elements remained intact apart from the end caps having been cut from three elements to facilitate their feeding into the input acid dissolution tank. The LAMA building was seriously damaged during the January 1991 bombing of Tuwaitha and, the plant components were salvaged and placed in temporary storage at the Al Shakili storage complex adjacent to the Tuwaitha site.

When it became clear that the project could no longer be housed in the LAMA building, the uranium recovery plant was redesigned – as Project 603 - in order that it could be re-installed at the Al Tarmiya site, which had sustained lesser bomb damage. Project 603 was to be limited to recovering the HEU from the fresh fuel elements and converting the recovered material in the form of UO_2. The UO_2 material was then to have been transferred to Project 247, where it would have been converted to UCl_4, in which form it could have been used as feed for EMIS separators to produce HEU.

The further enrichment of highly enriched uranium – Project 521C

It was planned to enrich further the uranium recovered from the irradiated HEU reactor fuel employing a 50 machine centrifuge cascade which was to be designed, fabricated and installed in Hall 9 of the EDC establishment at Rashdiya. The centrifuge machines were to be constructed partly from components already procured from foreign suppliers and partly from components ordered from Iraqi engineering companies.

The cascade was anticipated to include a mixture of centrifuge types, differing principally with regard to the rotor type – either carbon fibre or maraging steel. No attempts were made to assemble centrifuge machines from the available components, but if all the components required for the cascade were available, it would have been possible to assemble the machines at a rate of at least one per day. The basic civil engineering modifications were made to Hall 9, and concrete foundation strips were cast, on the existing floor, to accommodate a cascade of two parallel lines of 25 machines.

Although some shuttering had been assembled, none of the concrete mounting blocks for the centrifuge machines had been cast before the post-war decision was taken to abandon the project.

Not one single machine was completed for Project 521C, and consequently, no uranium was introduced into Hall 9.

Conversion to the metal of highly enriched uranium–Project 602/602B

Project 602 was designed to receive the recovered HEU from Project 601 in the form of UNH and to convert it to metallic form, which is the suitable feed material for casting the core components of the nuclear device.

The project was housed in Tuwaitha Building 64 and involved plant stages for the conversion of the input UNH through UO_4 to UO_2, the conversion of UO_2 to UF_4, the reduction of the UF_4 to uranium metal and systems for waste recovery. The stages for the conversion of the UNH to UO_4 were designed on the basis of laboratory-scale tests. They were fabricated, installed and commissioned using natural uranium feed.

The basic technology for the preparation of UF_4 was already well established, and an existing UF_4/uranium metal project, with a capacity of 20 kg uranium metal per day, designed around the end of 1989, was adopted for Project 602. This unit was installed, commissioned and produced a 10 kg test batch of natural UF_4 around the end of 1990. The reduction of UF_4 to uranium metal was well understood as the process had been in use for natural uranium since mid-1986. The principal development work required in this area was to further improve techniques in order to compensate for the process losses that would otherwise result from the small batch size of some 100 g that had been selected by the project managers. Although the waste recovery plant stages were not yet installed, the capability to commence HEU conversion from UNH to metal was essentially available in January 1991.

Building 64 was severely damaged in the January 1991 bombardment of Tuwaitha, and the project could no longer proceed in that building. The undamaged plant equipment was salvaged and stored. The project was redesigned and documented as Project 602B, but no practical measures were taken to reconstitute the capability.

The unit components that had been commissioned and thus contaminated with natural uranium were unilaterally destroyed. In contrast, other general-purpose components were retained for subsequent use in non-nuclear activities.

Miscellaneous Activities

Miscellaneous Activities Related to the Past INP

These include technical reports and projects that were partially or fully executed by various departments or groups attached to the former INP in support of the program but were essentially independent activities. They include the following:

1.   Abu-Skhair ore treatment/ Project 209 (Activity 2C).

This project was a pilot plant designed and constructed for the investigation of uranium ore processing from an Abu-Skhair mine to produce a yellow cake. The pilot plant design capacity was approximately 1000 kg/year of yellowcake.

Mining exploration tests for the deposit indicated serious technical and safety impacts on the mining works due to the effect of groundwater. The uranium mineralization was confined to a thin horizon of only 10-15 cm in thickness. The mine was closed and sealed under the supervision of the IAEA AT. A total of about 0.5 kg of the yellow cake was obtained from this ore. The pilot plant remained in its original location and was modified during 1992 to produce Alum from kaolin ore.

2. Purification of Oil Used in Diffusion Vacuum Pump

The project's purpose was to purify the waste oil from the diffusion pump connected to EMIS separators to permit recycling. The system was designed to treat 50 liters per batch. No detailed design was issued since the project was considered to be of low priority.

3. Assessment of Heavy Water Production Methods (Activity 2C)

This included a review of the standard separation methods of deuterium that will lead to heavy water production. No experimental or design work was carried out.

4. Chemical Cleaning of EMIS Separator Internals/ project 200 at Oqba General Establishment (OGE) (Activity 2C)

Internal parts of EMIS separators produced in OGE need to be degreased and chemically cleaned in order to attain a high vacuum and decrease the time for outgassing when used inside separators. This project was designed in Tuwaitha and executed (during 1989) in OGE. The project remained in operation and is currently used for cleaning equipment manufactured for the food industry and cleaning filters used in the power industry.

5. Transfer of Nuclear Fuel to Location B

During the aerial bombardment of the Tuwaitha site in Jan. 1991, it was decided to transfer all the fuel in the Tammuz-2 building outside of the Tuwaitha Site to prevent any possible catastrophic situation in case further aerial attacks on the Tammuz-2 reactor were mounted. The new location was chosen to be Jurf-al-Naddaf (location B). The total number of fuel elements stored in location B was 132 (two of which were natural uranium elements).

6. Disposal of Heavy water of Tammuz Reactors

In mid-1992, an order was issued to destroy 8 barrels and keep one. The heavy water of the barrels was disposed of into the Tigris river (near the guest house at Tuwaitha).

7. Production of CCl_4/ Project 214 (Activity 2C)

The objective of this project was to provide CCl_4 for the chlorination of UO_2 to produce UCl_4 and also to meet the needs for CCl_4 by other industries.

The basic design of a semi-continuous plant to produce 1000 tons/ yr was developed. No further steps were taken because the project was considered to be of low priority.

8. Work in Support of Project 411 (Activity 4G)

Vacuum plasma spraying of various parts used in the isotope separation chamber of project 411 was adopted to achieve a hardness higher than that normally obtained by heat treatment.

9. Purification of Secondary Explosives (Activity 40R)

A small laboratory building (15A) at Tuwaitha was equipped with simple apparatus for the recrystallisation of organic compounds to carry out R&D work to purify explosives. In April 1989, a decision was taken by PC3 to halt this activity and return all quantities of explosives to OGE.

10. Fuel Storage Project (Design & Project Directory)

During Jan. and Feb. 1991, the fuel storage areas near the reactors in Tuwaitha were hit and severely damaged. The fuel was transferred to a temporary storage area (Jurf Al-Naddaf) during Feb. 1991. In March 1991, it was decided to construct a more permanent fuel storage facility in Tuwaitha. The IAEA was informed of this decision in the summer of 1991, and the site was inspected by several inspection teams.

The execution of this project started in April 1991. When the earthworks were finished, the project was halted because the IAEA decided to remove all the fuel from Iraq to the Russian Federation.

Support for Non-Nuclear Groups and Activities

This includes work done by departments or groups within IAEC or PC3 supporting non-nuclear activities included in section (C) of UNSCR 687 (1991).

11. Production of Small Metallic Depleted Uranium Cylinders (Activity 4G) During 1988-1989, team 1 of Activity 4G received a request, issued from MIC, to produce small metallic depleted uranium cylinders for use (sub- calibre) kinetic energy slugs for test purposes. Natural uranium was used for the experiments. Rods of uranium metal of diameter 10-11 mm and 150 mm long was machined using a center lathe. They were heat-treated using a homemade tube furnace. The hardness achieved was 350-400 Hv. The cylinders were protected against oxidation by coating with nickel using the plasma vapour deposition unit at IAEC. About 100 pieces were made, of which about 40 were used by MIC, and the remainder were returned to IAEC, where they were dissolved in nitric acid and disposed of. The subject was not pursued further.

12. Production of Subcaliber from Metallic Uranium (Activity 4G)

During 1989 a request for sub-caliber production for antitank munitions was received from MIC. The produced experimental pieces (about 15) were made from natural uranium. Some were used in tests in a firing range and those not used were returned. Four of the returned were dissolved in nitric acid, and the remaining 5 pieces were handed over to the IAEA inspection team.

13. Special Inverter for Project 144/4 (Activity 2FE)

This work was concerned with designing and constructing a special power supply for a group of electrical loads comprising guidance and control systems for Al-Hussain missiles.

The prototype was built and passed the required quality control tests. At the end of 1990, six more power supplies were built and supplied to the user (Project 144/4).

The user agreed to purchase the equivalent military-grade components to build 200 such units. The purchase order was not executed.

14. Nuclear Battery for Satellite Communications (Activity 40CR)

The project aimed to separate strontium-90 from high-level radioactive waste to be used as an energy source for a nuclear battery. The project was adopted in Oct. 1990, but no experimental work was conducted on the project.

15. Production of Magnetic Materials for Project 144/4 (Activities 2A and 2F)

In March 1990, a request to produce permanent magnets was issued by Project 144/4. These were considered to be essential parts of guidance and control subsystems. As a result of the R&D effort, the production of magnetic materials of the AlNiCo type (C and ring shape permanent magnetics) was successful, and about 30 samples were produced and tested. The emphasis towards the end of 1990 shifted to VNiCo alloy, but no experiments were conducted. Two remaining samples were handed over to the IAEA on Feb. 16, 1997.

16. Work done in Al-Rabee Factory for Projects 144/4 and 1728

Samples (about 25) of graphite vanes with their bases for missile engines were machined from blocks from mid-1989 to mid-1990. Only one vane was coated and was handed to MIC. No further work was requested.

During the second half of 1990, injection head main rings (about 20) was machined.

17. Production of UDMH (Activity 2C)

Laboratory work was carried out from the early 1990 to produce UDMH. This work was conducted by Petroleum Research Center (PRC). Support in design and engineering was required from PC3. A preliminary PFD was prepared for processes reported in the literature. No basic design document was issued.

18. Material Analysis in Support of Projects 144/3 & 144/4 (IAEC, 2F and 40N) From 1987-1990 projects 144/3 and 144/4 received analytical support when samples were sent to IAEC then later to PC3 (less than 100 in all) for material analysis of stainless-steel alloys, aluminium alloys and carbon steel alloys.
19. Non-destructive testing of an RDX explosive (Activity 40B)

In June 1989, PC3 received a request from OGE to test the homogeneity of the chemical constituents along an RDX-filled cutting cord 30 cm long. The the test was carried out in the neutron radiography facility of the Tammuz-2 reactor.

20. Organizational Support for Al-Karama Project (PC3 head office and 2FE) MIC requested PC3 to study Al-Karama Project's organizational structure and suggest an optimal organizational structure. As a result, a complete re-structuring proposal was submitted to MIC on May 31, 1990.
21. Guidance and Control Support for Al-Karama Project (Activity 2FE)

A team was formed within PC3 during the third quarter of 1990 to suggest possible improvements on the missile circular error probability. There were no conclusive results by Jan. 1991.

22. Support to Al-Yawm Al-Azim Plant (Activity 2FE)

During the autumn of 1990, PC3 was requested to put into operation the missile fuel test station at the Al-Yawm Al-Azim plant.

After rectification, which took about one month, the test station was fully operational.

23. Al-Muntasir Project (Activity 2C)

A request was received from QGE in Nov. 1990 to produce basic and detailed design for this project. A basic design report was issued in Sep. 1991. However, no equipment was fabricated.

24. Project 3028 (Activity 2C)

A request was received from Al-Mutawakil Project (MIC) in mid-1990 to study the technical documents of an offer for a missile engine testing station, which used liquid fuel such as kerosene or hydrazine. The offer was from a Chinese company (CPMIEC). The document was studied, and an equipment list was prepared. No basic design was issued.

7. Status of documentation:

1. The IAEA possesses a complete set of IAEC/department 3000 and PC3 technical reports and documentation that were issued by the former INP. The number of technical reports issued was 1572. The Iraqi team (IT) provided titles, abstracts and descriptions of the contents of the five missing reports from the best recollection of their original authors. These missing reports covered the subject of minor importance to the former INP.
2. In addition to the technical reports, the IT handed over to the IAEA AT the following documentation related to the former INP, which should be available in its Iraq library:
    a. 1.5 million pages of technical and non-technical documents.
    b. Approximately 9 km of microfilm of technical and non-technical documents.
    c. A large amount of microfiches of documents.
    d. A very large holding of engineering design drawings and workshop drawings.
3. From this large documentation library, the IAEA AT has been able to assess and verify the full extent of the former INP and its temporal development.
4. The FFCD-F of approximately 2000 pages (including annexes) can only be viewed as a condensed summary of the work that was undertaken in the former INP.
5. Apart from FFCD-F, Iraq does not possess any of the technical documents that were issued by the former INP.
6. The missing technical documentation is in two categories as follows:

a. The first category concerns some missing drawings of gas centrifuge enrichment machines that were obtained by the Engineering Design Centre (EDC) during the late eighties through a foreign intermediary.

These drawings were produced by a German engineering company (well known to the IAEA).

These drawings covered two types of machines. The first being a single-cylinder sub-critical centrifuge, which was successfully tested by EDC and is fully described in the present document. The second centrifuge was a multi-cylinder super-critical machine of higher separative work output that was neither produced nor tested in Iraq. Some of these original drawings were handed over to the IAEA; others are missing and most probably destroyed. If the IT possesses any of these documents, then it is in its best interest to hand them over to the IAEA AT. It is illogical for IT to conceal any of these drawings. Moreover, if the IAEA AT wishes to review these drawings,, they should approach the government and/or company that originated them and request this review. In this manner, this matter could be easily resolved. In any case, the IAEA stated in paragraph 10 of the document (S/1997/127) dated Feb. 9, 1999, that *"the recovery of the missing drawings would do little to change the assessment of Iraq's capability in this area."*

b. The second type of missing documentation that is of concern to the IAEA AT is in the area of lens design that is part of the explosive package for a nuclear device. A few engineering drawings are missing and could not be located since the facility (Al-QaQaa) was bombed during the war in Jan/Feb 1991.

The design of these lenses is based upon the basic principles of detonation optics, and a simple computer code in two dimensions was written by the group at Al-QaQaa. This program was handed over to the IAEA inspection team on/29 during Oct. 17-24, 1995. The methodology adopted in this design was explained to IAEA inspection teams on many occasions. Also, some detailed questions were raised by the AT and discussed with the IT in the presence of Dr. P. Kurtz (an IAEA AT specialist from the Los Alamos Scientific Laboratory in the USA) on May 19, 1997. Clearly, the missing drawings were destroyed, and it is in the best interest of IT to hand them over to the IAEA AT had they existed. If the IAEA AT insists on obtaining these drawings,, the IT had suggested on several occasions during 1998 that the IT specialists concerned could reproduce these drawings under IAEA AT's supervision using the original computer code in IAEA's possession. This could be one way of resolving this matter completely. In any event, the IAEA dealt with this question and stated in paragraph 9 of document S/1999/127 dated Feb. 9, 1999, that *"To compensate for this uncertainty the OMV plan is predicated upon the assumption that Iraq has acquired the technical ability to fabricate an implosion-type fission-based nuclear weapon."*

8. Declarations of the Past Iraqi Nuclear Program (INP)

Early Declaration

During the period May 1991 - July 1992, the Iraqi Team (IT) handed over to the IAEA Action Team (AT) a large number of documents and technical reports describing all activities of the former Iraqi Nuclear Program (INP). In addition, the IT took the initiative. It submitted a 60-page document that amounted to a brief description that highlighted major milestones of the former INP. This document was to be read in conjunction with other information provided to the IAEA by the IT. According to information from the IAEA AT, approximately 150 experts from various member states reviewed this document, and 24 comments were received. The IT responded to these comments and issued a new version of the document of about 100 pages that were submitted to the IAEA near the end of 1992.

The Full, Final and Complete Declaration (FFCD-F)

1. In its letter dated Nov. 20, 1995, the IAEA AT requested the IT to prepare a stand-alone FFCD document describing the former INP that could be understood without the need to refer to the large amount of supplementary documents and reports already available to the IAEA.

2. A preliminary list of contents of the proposed FFCD was handed over by the IT to the deputy AT leader (Mr. G. Dillon) early in Dec. 1995, and AT comments on the list of contents were invited. Some AT comments were received, and the preliminary list of FFCD contents was modified accordingly. The IAEA submitted the agreed list of contents as an annex to its report to the Security Council (document S/1996/261) on April 11, 1996. The Member States of the Security Council (including the permanent five) approved the report and this list of contents.

3. The IT responded speedily to the IAEA's request and submitted a draft FFCD document of six volumes comprising more than 1000 pages on March 1,1996. The difficulties faced by the IT in preparing this document were due to the following:

   a. The contents were mostly based on the best recollection of the Iraqi personnel that were selected for drafting the various sections since all relevant former INP documents had been previously handed over by the IT to the various IAEA inspection teams.

   b. The events in question, that were to be written up from memory, were more than five years old at that time.

4. Discussions of the draft FFCD began belatedly, more than ten weeks after submitting the draft document, on May 13, 1996, with the Thirtieth.1 IAEA team. As a result of these discussions,, the IT draft was revised and submitted to the IAEA AT on June 22, 1996.

5. This second draft was discussed with an IAEA action team in the presence of the AT leader (the late Prof. M. Zifferero) during the period June 24-27, 1996. Also, the deputy AT leader at that time (Mr. G. Dillon) made numerous useful editorial comments on this draft, the last of which was received on Aug. 5, 1996.

6. The final version of the FFCD (labeled FFCD-F) of more than 1600 pages was handed over to the IAEA AT leader on Sept. 7, 1996.

7. After a torpid four months, the AT presented the IT with (42) comments on Jan. 13, 1997, (17) comments on Feb. 5, 1997, and then a further (12) comments on Feb. 12, 1997.

8. These belated comments were responded to speedily by the IT, who submitted a full 30-page reply as an annex to FFCD-F on Feb. 26, 1997.

9. As foreseen by the IT, the IAEA in its April 8, 1997 report to the UN Security Council (document S/1997/ 297), claimed that "Iraq provided a written response to the matters discussed. This response is currently under evaluation etc.". Indeed, in this way, the AT evaded stating in their sixth monthly report to the UNSC that "Iraq has completed all actions contemplated in UNSCR 687 (1991) in the nuclear area".

10. The additions and revisions to the FFCD-F, which were sent to the IAEA AT on Feb. 26, 1997, were discussed belatedly during an IAEA inspection team visit to Iraq (Thirtieth.4) during the period May 16-22, 1997.

11. Following the May discussions, the IAEA AT sent a number of comments dealing with minor aspects of the FFCD-F. All these matters were responded to in a series of letters during July 1997, which were further discussed during an IAEA inspection team visit to Iraq during the period July 19-24, 1997 (Thirtieth.5). all the comments arising from the July discussions were answered to the satisfaction of the IAEA on August 3, 1997. A list of these letters is given in attachment 8.2. The IAEA AT, in its letter of August 6, 1997, congratulated the IT on the completion of these hopefully "final" amendments.

12. In the opinion of the IT, the FFCD-F together with its annexes is full, final and complete to the point that the DG of the IAEA made the following remark in his introductory statement to the Board of Governors on June 9, 1997: *"As I have previously reported the Agency has for some time been at a point of diminishing returns with respect to furthering our knowledge of the details of Iraq's past nuclear program."*

13. The IAEA stated in para 79 of its fourth consolidated report to the Security Council (S/1997/779) submitted on Oct. 8, 1997, that:

"There are no significant discrepancies between the technically coherent picture which has evolved of Iraq's past program and the information contained in Iraq's FFCD-F issued on 7 September 1996 as supplemented by the written revisions and additions provided by Iraq since that time."

14. It can safely be concluded that FFCD-F was issued on Sept. 7, 1996. as supplemented by the written revisions and additions provided by the Iraqi team was accepted fully by the IAEA, and the Security Council was informed accordingly, through the fourth IAEA consolidated report was submitted on Oct 8, 1997.

8.1 Currently Accurate, Full and Complete Declaration (CAFCD)

1. On Nov. 8, 2002, the Security Council adopted resolution S/RES/1441 (2002).
2. Iraq agreed to deal with this resolution.
3. The present document presented here is a currently accurate, full and complete declaration (CAFCD) of all aspects of its nuclear program. It has been prepared to fulfill Iraq's obligations in accordance with paragraph 3 of UNSCR 1441(2002). The contents of the CAFCD are listed in attachment 8.2.

9. Present Status of Facilities, Equipment and Materials and Nuclear Material Balance Status of Facilities, Equipment and Materials

1. Table 9.1.1 lists the main buildings of the former INP that were either partially or totally destroyed by aerial bombardment and missile attacks during the war in Jan. - Feb., 1991. Those buildings that were partially destroyed were later completely destroyed under IAEA AT supervision, and these are noted in the above table.
2. Table 9.1.2 lists the main buildings that survived the war of Jan. - Feb., 1991 but were later destroyed under IAEA AT supervision during the period April 1992-November 1993.
3. Table 9.1.3 is a summary of the main equipment and materials of the former INP that were destroyed or rendered harmless under

IAEA AT supervision. The total quantity of items that were included in this category exceeded 2800.

4. Table 9.1.4 lists the main equipment and materials of the former INP that were removed and transported by the IAEA AT outside of Iraq. A few items were specified by the IAEA AT in May 1997 for the purpose of verification, and these, were located by the IT and handed over to the IAEA AT in July 1997, are also included in this Table.

5. UNSCR 687 (1991) stated that *"Iraq shall unconditionally agree not to acquire or develop nuclear weapons or nuclear weapons-usable material or any sub-systems or components or any research, development, support or manufacturing facilities related to the above."* Based on this resolution, all buildings, systems, components, subsystems and other items specified in para 12 of resolution 687 were destroyed, removed or rendered harmless either unilaterally by Iraq and verified by the AT or directly by the IAEA AT. In addition, a considerable number of items were destroyed during the war. Dual-use items were tagged as stipulated under the OMV plan. The report of the seventh inspection team in Oct. 1991 stated that *"a large amount of EMIS and centrifuge equipment was destroyed during the seventh inspection"*. The destruction process was continued by the eighth inspection team, whose report stated *"Activities during the seventh inspection, to destroy or render harmless equipment components associated with the Iraqi uranium enrichment program continued during the eighth inspection."* In addition, the reports of the eleventh and twelfth inspection teams in 1992 stated that *"the destruction of the buildings and equipment at the Al-Atheer - Al- Hatteen site has been completed."* Also, the fifteenth inspection team in Nov. 1992 destroyed a number of additional equipment and materials. On Oct. 19, 1996, the IT handed over to the IAEA AT a document that contained detailed information on the status, locations, and quantities of main equipment, including those still outside Iraq due to the embargo imposed in August 1990. Moreover, the Iraqi team formally authorized the IAEA AT to dispose of what it saw appropriate of these equipment and materials belonging to the

former gas centrifuge program. In this respect, during the period May 23-25, 2000, the IAEA AT completed the destruction, in Jordan, of a filament winding machine and its associated spare parts and raw materials that was owned by the former INP (see paragraph 17 of IAEA document GOV/2000/35-GC(44)/11 for more details).

6. As part of the FFCD-F and its supplementary attachments, the IT submitted more details on the status of equipment and materials, including their storage locations, movements, and destruction sites. The IAEA AT directly verified their status. After removing the main equipment from destruction pits to storage, the eighth IAEA inspection team requested the IT cover all destruction pits. Also, the twelfth IAEA inspection team, and after the destruction of Al-Atheer, agreed to the removal and transfer of all debris. During 1997, the IAEA AT requested all destruction pits to be uncovered in order to verify the remains of the destroyed equipment and materials. The IT submitted to the IAEA AT on Feb. 17, 1997, a report on the status of equipment comprising 78 pages. This report also included adequate clarifications to all IAEA AT queries and comments concerning the already destroyed equipment of the former INP.

7. The status of equipment and materials remained unchanged since 1991 except for the following:

    a. Labeling the equipment according to the IAEA AT code and submitting a new version of the inventory.
    b. Changing equipment locations as mentioned in FFCD-F attachments.
    c. Verification of equipment movement and analysis of missing items. The Iraqi team assisted the IAEA AT in this respect by assigning personnel to accompany the IAEA AT members during their many visits to the destruction sites.

8. For verification purposes, the IT submitted on July 8, 1997, to the IAEA AT an inventory of 62 pages of destroyed equipment and materials that have been identified and fully accounted for.

9. In response to an IAEA AT request, the IT submitted a new version of addendum I part 2 annex 1 of the FFCD-F on Aug. 9, 1997. This addendum dealt with the movement and destruction of equipment and materials.

In conclusion, all relevant facilities, equipment and materials of the former INP were destroyed or rendered harmless. Therefore, Iraq no longer possesses any material capability in the nuclear field.

10. Para 72 of the fourth IAEA consolidated report (S/1997/779) to the Security Council dated Oct. 8, 1997, stated the following:

"All known indigenous facilities capable of producing amounts of uranium compounds useful to a reconstituted nuclear programme have been destroyed along with their principal equipment.

All known procured uranium compounds are in the custody of the IAEA.

All known practically recoverable amounts of indigenously produced uranium compounds are in the custody of the IAEA.

All known facilities for the industrial-scale production of pure uranium compounds suitable for fuel fabrication or isotopic enrichment have been destroyed, along with their principal equipment. All known single-use equipment used in the research and development of nrichment technologies has been destroyed, removed or rendered harmless. All known dual-use equipment used in the research and development of enrichment technologies is subjected to ongoing monitoring and verification.

All known facilities and equipment for the enrichment of uranium through EMIS technologies have been destroyed along with their principal equipment.

IAEA inspections have revealed no indications that Iraq's indigenous plutonium production reactor plans proceeded beyond a feasibility study.

The facility used for research and development of irradiated fuel reprocessing technology was destroyed in the bombardment of Tuwaitha, and the process-dedicated equipment has been destroyed or rendered harmless.

The principal building of the Al-Atheer nuclear weapons development and production plant has been destroyed, and all known purpose-specific equipment has been destroyed, removed or rendered harmless.

The entire inventory of research reactor fuel was verified and accounted for by the IAEA and maintained under IAEA custody until it was removed from Iraq".

Table 9.1.1
List of main buildings of the former INP destroyed during
the aerial bombardment (January - February 1991)

Site	Destroyed Buildings
Tuwaitha Nuclear Research Centre	- Radiochemistry Laboratories (Bldg. 9)
	- Physics Department (Bldg. 10B)
	- Laboratory for uranium Metal Preparation (Bldg. 10)[1]
	- IRT-5000 Reactor (Bldg. 13)
	- Computer Hall and Offices (Bldg. 13 part)
	- Electrical Sub-Stations (Bldg. 14, 72, 84)
	- Radioisotope Production Department (Bldg. 15A)[1]
	- Quality Control of Radioisotope Production Department (Bldg. 15B)[1]
	- LAMA Laboratories (Reprocessing), Bldg. 22)
	- Experimental Workshop, laser and Plasma Studies (Bldg. 23)[1]
	- Tammuz-2 Reactor (Bldg. 24)
	- Store and workshop (Bldg. 28)
	- Decontamination Laboratory (Bldg. 27)
	- Chemical Coating Laboratory (Bldg. 30)
	- Cooling Tower for Tammuz-2 Reactor (Bldg. 31)
	- Radioactive Waste Treatment Station (RWTS, Bldg. 35)
	- Calibration Laboratories and Decontamination Area (Bldg. 41)
	- Laboratories for Material Processing (Bldg. 83)
	- Laboratories for Uranium Treatment and Liquid Radioactive Waste (Bldg. 64)
	- Laboratories for Experimental Physics and Measurements (Bldg. 68)
	- Hydrogen Station (Bldg. 70)
	- Sewage Station for 30 July Project (Bldg. 71)
	- Experimental Research Laboratories for Fuel Fabrication (Bldg. 73 complex)[1]
	- Cooling Tower of bldg. 80 (Bldg. 79)

	- Laboratories for EMIS Development (Bldg. 80)[1]
	- Laboratories of UCl_4 Preparation and Purification (Bldg. 85)[1,2]
	- Chemical Enrichment Laboratories (Bldg. 90)
	- Mechanical workshop (Bldg. 57)
	- Material studies (Bldg. 63)
	- Electrical engineering design labs. (Bldg. 82)
Al-Atheer Centre	- High Explosives Test Bunker and stores (Bldg. 33)[2]
	- Offices of Activity 40B (bldg. 79)
	- Electrical Laboratories (Bldg. 94)
Tarmiya EMIS Facility	- EMIS stage 1 Separator Building (Bldg. 33)
	- Air Conditioning Units (Bldgs. 21-23, 94-96, 244, 248)
	- EMIS stage 2 Separator Building (Bldg. 245)
	- Electrical power Sub-Stations (Bldgs. 5, 38, B1, 243, 228)[2]
	- EMIS Separator Wash Room (Bldg. 225)[2]
	- Waste Treatment Building (Bldg. 216)
	- Chemical process building (Bldg. 210)
	- Chemical process building (Bldg. 230)
Sharqat EMIS Facility	- EMIS Washing and Cleaning (Bldg. C-034)
	- EMIS Washing (C-054)
	- Electrical Power Supply (Bldgs. B-029, B-027, B-020, B-032)[2]
	- Utility Building (Bldg. B-031)
	- Cooling Towers (Bldg. B-033)
	- Equipment Hall (Bldg. B-051)
	- Main Power Station (B-046)
	- EMIS stage 1 Separator Hall (B-021)
	- Workshop (B-003)
Al Qaim Uranium Purification Facility	- Uranium Purification Building (Bldg. 300)
Jazira Uranium Processing Plant	- UO_2 Production Plant (Blgd. 000)

	- UCl$_4$ Production Plant (Bldg. 400)
	- UCl$_4$ Production Plant Utilities
	- UO$_2$ Production Plant Utilities

1. Iraq further levelled buildings to the ground.
2. Building further destroyed under IAEA supervision.

Table 9.1.2
List of main buildings of the sites of the former INP that survived the war but were later destroyed under IAEA supervision

Destruction Date	Site	Destroyed Buildings	Destruction Method
April- May 1992, IAEA-11/12	Al-Atheer Centre	- Carbide (uranium machining), Bldg. 55 - Casting (uranium metallurgy), Bldg. 50 - Quality Control. Bldg. 19 - Explosion Chamber, Bldg. 1B (cutting with torches) - High Explosives Test Bunker, Bldg. 33 - Physics (gas gun), Bldg. 21 - Polymer (uranium metal processing), Bldg. 84 - Powder Preparation, Bldg. 82	Demolition with explosives. Bldg. 33 was filled with concrete and scrap metal; the protective berm has been removed
July -September 1992, IAEA-13/14	Tarmiya EMIS Facility	- Electrical Sub-Stations, Bldg. 5, 38, 243 - EMIS stage 2 Separator Building, Bldg. 245	Demolition with explosives/heavy machinery
July -September 1992, IAEA-19/14	Al-Sharqat EMIS Facility	- Electrical Sub-Stations, Bldgs. B- 20,B-27, B-29 - EMIS stage 2 Separator Building, Bldg. B-21	Demolition with explosives/heavy machinery
November 1993, IAEA-22	Abu Skhair Mine	Abu Skhair uranium mine	Backfilled, shaft door welded and sealed

Note: Electrical power supplies to the Tarmiya and Al-Sharqat sites were reduced by order of magnitude.

Table 9.1.3
Main equipment and materials of the former INP which were destroyed or rendered harmless under IAEA Supervision

Time Period	Program Activity	Equipment Location	Main Components	Destruction Method	Total Quantity
Oct. -Nov. 1991,IAEA7/8	Gas centrifuge enrichment	Engineering Design Centre, Al-Furat Centrifuge Production Facility	All detectedcentrifuge components and important related equipment were eitherremovedbythe Inspection Teams, rendered harmless or destroyed, including: Centrifuge housings, aluminum rotor tubes, carbon fibre cylinders, complete rotor assemblies, unfinishedmaragingsteel cylinders, molecular pumps, motor stators, bearings,frequency convertors, balancing machine, centrifugetest jigs, complete oil centrifuge, oil centrifuge cylinders, centrifuge jackets, UF6feedingsystem, miscellaneous parts of the machine tools, AlNiCoand SmCoring magnets,jigforMIG welder, mandrel forflowformingmachine, electron beam welder fixture, rotating spindle andmandrelfor CNC machine tool, special collect and whirling head, specific fixtures for centrifuge production.	Mainly by crushing or flame cutting	More than1790 components and items

Electromagnetic Isotope Separation (EMIS)	Tarmiya EMIS Facility, Tuwaitha Nuclear Research Centre, Daura (SEHEE), Amin(Um Al-Maarik)	Vacuum chambers, coils, Collectors, injector power supply, ion sources, ironsystems, poles, coil-windingmachines, andelements ofmachine tools.	Mainly by flame cutting	More than340items
Reprocessing	Tuwaitha Nuclear Research Centre	Chopping machine, glove boxes, manipulators, cables for manipulators, mixer settlers, hot cells, dissolver	The glove boxes were filled with cement. Mixer settlers were filledwithepoxyresin. Hotcells, dissolver and chopping machine were rendered harmless by cutting and removal of manipulators.	More than40ite
Chemical Isotope separator	Tuwaitha Nuclear Research Centre	Glass columns(10)and other items used in the chemical separation work.	Smashed	More than10items

Jan. 1992, IAEA-9	Gas centrifuge enrichment	Engineering Design Centre, Al-Furat Centrifuge production facility	Aluminum alloys in the form of tube extrusions (morethan500 tones),ferrite magnets, aluminiumupper flange forgings (9.000), aluminium jacket ring forgings(9.000), bottom flange(250).	Melting and mixing with lower grade materials. Ferrite magnets were destroyed by crushing.	More than500tones of materials

Table 9.1.3 (continued -1)

Time Period	Program Activity	Equipment Location	Main Components	Destruction Method	Total Quantity
Apr. -May 1992, IAEA-11/12	Device development	Al-Atheer Centre	Cold and hot isostatic presses, furnaces, plasma spray systems, machine tools, vacuum pumps, power supplies	Flame cutting, destruction with explosives	more than 50 items
April and November 1992, IAEA-11/15	EMIS	Tarmiya EMIS Facility, Tuwaitha Nuclear Research centre	Experimental EMIS magnet system with 9 double poles, coil winding machines and their accessories, HEPA filter elements and exhaust air filtration units	Mainly by flame cutting. Filtration components were crushed	More than 10 items and 285 filter elements and units
November 1992, IAEA-15	Gas centrifuge enrichment	Engineering Design Centre	350-grade maraging steel tools and cylinders unilaterally destroyed by Iraq (761) were further adulterated byre-melting and diluting it with equal amounts of high carbon steel in Basra Foundry	Melting and mixing with lower grade materials	About 76 tones of maraging steel

Note: Many items of the equipment of the former INP were destroyed in the aerial bombardment (January - February 1991) and were confirmed by the IAEA as not recoverable or rendered harmless

Table 9.1.4

Main equipment and materials of the former INP removed by IAEA outside of Iraq

Time Period	Program Activity	Equipment Location	Examples	Total Quantity
From October 1991 to April 1992, IAEA-7/8/9/11	Gas centrifuge enrichment, Device development, radiochemistry	Tuwaitha Nuclear Research Centre, Engineering Design Centre, Al-Atheer Centre	Examples of major centrifuge components (rotor tubes, end caps, pin bearings, etc.), centrifuge rotors, HEPA air filters, computer codes, high-speed streak video cameras and their components, holding collar for the mandrel, beryllium metal, flow-forming roller, die used to manufacture the explosive lenses, parts of the CNC coordinate measurement machine.	More than 200 items
From September 1995 to April 1996, IAEA-28/29	Gas centrifuge enrichment, Device development	Tuwaitha Nuclear Research Centre, Engineering Design Centre, Al-Atheer Centre, Al- Qa Qaa GE	AlNiCo and CoSm ring magnets, maraging steel (17 tones), spools of high modulus and high tensile strength carbon fibres, cylindrical initiator, thermal batteries, wavefront shape measurement device, tape with back-up of the 1 computer codes, 32-point electrical distributor for firing set, detonators and 1 1 1 ionisation probes, krytrons,8-channel ionisation probe analyser card .	More than 20 items and more than 200 ring magnets

(1) These items were specified by the IAEA AT for verification in May 1997 and were located by the IT and handed over to the AT in July 1997

.2. Nuclear Material Balance:

1. In accordance with para 12 of UNSCR 687 (1991), Iraq declared its nuclear materials. It placed them under the exclusive control and custody of the IAEA since April 1991. Iraq had 539.5 tons of uranium compounds as yellowcake or refined UO_2 and as follows:
    a.   338 tons of imported yellowcake.
    b.   6 tons of imported depleted uranium.
    c.   27.5 tons of imported UO2, part of this had been used within the former INP.
    d.   168 tons of yellowcake that were produced indigenously by extraction from phosphoric acid in Akashat and were subsequently converted to UO2 at Al- Jazira site. This nuclear material was collected from different locations in Iraq and inventoried by IAEA inspection teams. Then, after verification by the IAEA, this material was placed under its custody in a single location near Baghdad.

2. In addition, Iraq had 145 kg of uranium as reactor fuel for the (14) and (17) July research reactors in various enrichments and in two categories (fresh and spent fuel) as shown in table 9.2.1. This fuel was under IAEA safeguards before 1991 and remained so thereafter. The IAEA AT verified the disposition of this fuel; then it was removed from Iraq in three shipments: the first was late in 1991, which included all the fresh fuel, and the last two shipments were in Dec. 1993 and Feb. 1994, which included all the spent fuel. Table 9.2.2 shows the total amount of plutonium (<5g) that was removed from Iraq by the IAEA.

The IAEA AT efforts in the verification of all nuclear materials (quantitative and qualitative) started when the first IAEA inspection team visited Iraq in May 1991 and continued until Sept. 1996. This verification process included physical and chemical assays (destructive

and non-destructive) covering all nuclear materials, including those that were not covered by IAEA safeguards (e.g. yellow cake and various wastes).

3. The verification activities implemented by IAEA eventually confirmed the Iraqi declaration of July 7, 1991, concerning the inventory of nuclear materials.

4. The DG of the IAEA stated in his briefing to the UNSC on Nov. 7, 1996, that *"No further objects in the nuclear sphere have been the subject of destruction, removal or rendering harmless since the last consignment of highly enriched uranium was air-lifted to Russia in February 1994".*

5. The IT responded to some AT remarks on the question of nuclear material balance in a letter dated Sept. 10, 1996, that included final clarifications. Since that date, the IT did not receive any further remarks of any substance from the IAEA, which indicates that the balance of nuclear materials is final and complete, thereby implying that Iraq had adequately fulfilled its obligations in this important area.

6. Paragraph 72 of IAEA's report to the UNSC (document S/1997/779 dated Oct. 8, 1997) stated the following:

"All known procured uranium compounds are in the custody of the IAEA.

All known practically recoverable amounts of indigenously produced uranium compounds are in the custody of the IAEA.

The entire inventory of research reactor fuel was verified and accounted for by the IAEA and maintained under IAEA custody until it was removed from Iraq".

Table 9.2.1
Uranium Fuel Removed From Iraq IAEA Supervision

No.	Date of Removal	Element Weight (g)	U-235 Weight (g)	No of Items	Uranium enrichment (%)	Irrad. Status
1.	17-11-1991	13722	10998	68	80	Fresh
2.	17-11-1991	3538	1272	10	36	Fresh
3.	04-12-1993	86480	8648	68	10	Irrad.
4.	04-12-1993	1002	360	3	36	Irrad.
5.	12-02-1994	8150	6588	41	80	Irrad.
6.	12-02-1994	1280	128	1	10	Irrad.
7.	12-02-1994	11041	8872	55	80	Irrad.
8.	12-02-1994	11874	11050	38	93	Irrad.
9.	12-02-1994	7900	55	2	Natural	Irrad.
	Total	144987	47971	286		

Note: 1- Uranium fuel was transferred to the Russian Federation. 2- In November 1991, IAEA also removed 63 mg of U 233.

3- Uranium fresh fuel components of 323 g (36% enrichment) were exempted by Iraq from safeguard and 417 g (93% enrichment) to the IAEA Seibersdorf Laboratory.

Table 9.2.2
Plutonium Removed From Iraq under IAEA Supervision

No.	Date	Weight	Plutonium Isotope	No of Items	Origin
1.	June 1991 IAEA-2	< 5g	Pu		Iraq
2.	October 1991 IAEA-7	Microgram quantities	Pu-238	2 items	Iraq
3.	November 1991 IAEA-8	Milligram quantities	Pu-239	6 sealed ampoules	Amersham, UK
4.	November 1991 IAEA-8	Microgram quantities	Pu-238	33 sealed ampoules	Amersham, UK
5.	November 1991 IAEA-8	< 0.3 g	Pu		Iraq

Note: 1- Plutonium was transferred to the IAEA Seibersdorf Laboratory.
2- Two Np-237 standards (about 200 mg) were also removed by IAEA (November 1991).

10. IAEA Activities in Iraq during May 1991- Dec. 1998

1. Table 10.1 shows that 34 IAEA inspection teams visited Iraq during the period May 1991 until July 1997. The average number of IAEA inspectors per team was 18 and their average duration of stay in Iraq was 8 days. They carried out

512 inspections of various nuclear-related and unrelated sites (excluding inspections carried out under the Ongoing Monitoring and Verification (OMV) plan). The overall inspection effort expended in Iraq (excluding OMV inspections) was 5274 man. days. During the same period, the Director-General of the IAEA submitted 44 reports to the UNSC. **There were no IAEA inspection activities in Iraq since July 1997 other than those carried out under the OMV plan.** Table 10.2 shows details of the IAEA AT delegations that visited Iraq between Feb.-Dec., 1998 for various discussions with the Iraqi Team (IT).

2. Upon the request of the IAEA on November 20, 1995, the Iraqi team (IT) submitted a draft of the "Full, Final, and Complete Declaration (FFCD)" of its former nuclear program on March 1, 1996. After two discussion sessions with IAEA teams, the final document (FFCD - version F) was submitted on September 7, 1996. Upon further request of the IAEA Action Team (AT), the IT added many detailed clarifications that were included as annexes to FFCD- F, the last of which was a slight amendment submitted on August 3, 1997. The activities associated with this effort are described in section 3 of this summary. Para 79 of the fourth consolidated report of the Director-General of the IAEA (S/1997/779), to the Security Council dated Oct. 8, 1997, stated:

"There are no significant discrepancies between the technically coherent picture which has evolved of Iraq's past programme and the information contained in Iraq's FFCD-F issued on 7 September 1996 as supplemented by the written revisions and additions provided by Iraq since that time."

3. Implementation of the ongoing monitoring and verification (OMV) plan pursuant to UNSCR 715(1991) has been up and running since 1994. **Table**

10.3 shows details of these activities; no proscribed equipment, materials, or activities exist in Iraq, and more than 2154 IAEA NMG (no notice) inspections have been carried out since 1994 to confirm this fact. The IAEA unilaterally suspended its OMV activities in Iraq on Oct. 29, 1997. The IAEA also decided unilaterally to evacuate its staff from Iraq on Nov. 13, 1997, and repatriated them to Baghdad to resume their normal activities on Nov. 21, 1997, and then ceased all OMV activities after Dec. 15, 1998, when the resident NMG group departed from Iraq at the behest of the executive chairman of UNSCOM and without the approval of the UNSC.

4. The IAEA (AT) is required to report that "Iraq has completed all actions contemplated in UNSCR 687 (1991), which paves

the way, together with a similar report by UNSCOM, to lifting the embargo, i.e. the implementation of para 22 of the same resolution" claimed in para 75 of the document S/1997/779 to have insufficient information on some minor subjects. The IT fully responded to the requirements of the IAEA AT and, in a series of 38 letters listed in table 8.2. **Discussions of these minor aspects and any new points that may arise in the future could be carried out as part of the OMV activities as suggested by the IAEA in para 83 of their report S/1997/779.** Section 10 of this summary presents an updated status of these minor points.

5. During the meetings held in Vienna, July 3-5, 2002, between the Iraqi delegation and the delegation of the Secretary-General of the United Nations, the Director-General of the IAEA reiterated the position of the IAEA in that: disarmament phase in the nuclear area has been completed. The remaining three points were considered by him to be of minor importance and could be resolved during OMV. The Iraqi delegation stated that no proscribed activity was undertaken since Dec. 1998.

Table 10.1
IAEA inspection activities in Iraq during the
period May 1991- December 1998

No.	Team	No. of inspectors	Period	Man. Days	No. of Facilities Inspected
1.	First	34	May 15-21, 1991	204	7
2.	Second	18	June 22-July 3, 1991	198	7
3.	Third	37	July 7-19, 1991	407	15
4.	Fourth	20	July 27-Aug.10, 1991	280	22
5.	Fifth	15	Sept.14-20, 1991	90	3
6.	Sixth	44	Sept.22-30, 1991	352	6
7.	Seventh	39	Oct.11-22, 1991	429	18
8.	Eighth	19	Nov. 11-18, 1991	133	10
9.	Ninth	14	Jan.11-14, 1992	42	5
10.	Tenth	31	Feb.5-13, 1992	248	19
11.	Eleventh	26	April 7-15, 1992	208	17
12.	Twelfth	27	May 26-Jun 4, 1992	243	23
13.	Thirteenth	9	July 14-21, 1992	63	5
14.	Fourteenth	15	Aug.31-Sept. 7, 1992	105	11
15.	Fifteenth	38	Nov. 8-18, 1992	380	29
16.	Sixteenth	8	Dec. 5-8, 1992	24	3
17.	Seventeenth	8	Jan. 25-31, 1993	48	10
18.	Eighteenth	23	March 3-11, 1993	184	35
19.	Nineteenth	14	April 30-May 7, 1993	98	33
20.	Twentieth	10	June 25-30, 1993	50	10
21.	Twenty-First	16	July 24-27, 1993	48	21
22.	Twenty-Second	17	Nov. 1-15, 1993	238	41
23.	Twenty-Third	17	Feb. 4-11, 1994	119	41
24.	Twenty-Fourth	15	April 11-22, 1994	165	39
25.	Twenty-Fifth	12	June 22-July 1, 1994	108	24
26.	Twenty-Sixth	18	Aug. 22-Sept. 9, 1994	288	16
27.	Twenty-Seventh	8	Oct. 14-21, 1994	56	30
28.	Twenty-Eighth	15	Sept. 9-20, 1995	165	5

29.	Twenty-Ninth	13	Oct. 17-24, 1995	91	3
30.	Thirtieth.1	12	May 13-19, 1996	72	0
31.	Thirtieth.2	4	June 24-29, 1996	20	1
32.	Thirtieth.3	3	Feb. 5-7, 1997	6	2
33.	Thirtieth.4	12	May 16-22, 1997	72	0
34.	Thirtieth.5	8	July 19-24, 1997	40	1
	Total	619		5274	512

Table 10.2
IAEA AT delegations deputed to Iraq during the period Feb. – Dec. 1998

No.	Team	No. of delegation members	Period	Man. Days	No. of facilities visited
1.	IAEA-DEL	4	Feb. 14-22, 1998	32	0
2.	IAEA-DEL	1	March 22-April 3, 1998	12	1
3.	IAEA-DEL	3	March 26-April 1, 1998	21	0
4.	IAEA-DEL	3	June 1-4, 1998	9	4
5.	IAEA-DEL	3	June 29-July 3, 1998	12	0
6.	IAEA-DEL	3	Sept. 14-19, 1998	15	0
7.	IAEA-DEL	3	Dec. 9-12, 1998	9	0
	Total	20		110	5

Table 10.3
IAEA Nuclear Monitoring Groups in Iraq
During the Period Sep. 1994-Dec. 1998

No.	Team	Period	No. of inspectors	No. of Visits	No. of facilities inspected	
					Nuclear related sites	Nuclear unrelated sites
1		Sept. 7-28, 1994	4	21	17	0
2		Sept. 29 – Oct. 20, 1994	2	16	14	0
3		Oct.21-Nov. 7, 1994	3	18	7	0
4		Nov. 8-28, 1994	5	13	9	0
5		Nov. 29-Dec. 15, 1994	3	13	13	0
6		Dec.16,1994-Jan.11, 1995	3	24	12	6
7		Jan.12-Feb.1, 1995	4	17	12	2
8		Feb. 2-27, 1995	3	19	16	1
9		Feb. 28-March 15, 1995	4	9	7	2
10		March 16-April 5, 1995	5	22	11	11
11		April 6-26, 1995	6	30	21	8
12		April 27-May 10, 1995	5	14	7	0
13		May 11-30, 1995	3	16	13	0
14		May 31-June 20, 1995	4	19	12	0
15		June 21-July 9, 1995	4	16	14	0
16		July 10-30, 1995	4	20	15	2
17		July 31-Aug. 10, 1995	4	6	4	1
18		Aug.11-29, 1995	5	18	13	1
19		Aug. 30-Sept. 11, 1995	2	11	11	0
20		Sept. 12-Oct. 3, 1995	3	16	15	0
21		Oct. 4-21, 1995	4	46	12	16
22		Oct. 22-Nov. 8, 1995	3	12	11	0
23		Nov. 9-19, 1995	3	12	9	0
24		Nov. 20-Dec.11, 1995	10	30	24	1
25		Dec.12, 1995-Jan.3, 1996	9	24	18	0
26		Jan. 4-26, 1996	4	20	12	0
27		Jan. 27-Feb.11, 1996	7	16	14	0

28		Feb. 12-March 4, 1996	3	16	14	0
29		March 5-25, 1996	5	18	5	1
30		March 26-April 15, 1996	4	18	14	0
31		April 16-May 5, 1996	5	30	12	17
32		May 6-28, 1996	5	31	25	2
33		May 29-June 17, 1996	4	11	2	5
34		June 18-July 8, 1996	7	13	6	1
35		July 9-29, 1996	4	21	20	0
36		July 30 – Aug. 19, 1996	4	21	19	1
37		Aug. 20- Sept. 15, 1996	5	24	21	2
38		Sept. 16 – Oct. 5, 1996	4	12	12	0
39		Oct. 6-21, 1996	6	31	11	19
40		Oct. 22 – Nov. 10, 1996	6	34	24	3
41		Nov. 11-Dec. 1, 1996	6	20	11	3
42		Dec. 2-18, 1996	4	20	14	2
43		Dec. 19, 1996, Jan. 5, 1997	3	8	7	1
44		Jan. 6-29, 1997	4	28	23	4
45		Jan. 30-Feb. 19, 1997	3	15	12	2
46		Feb. 20-March 12, 1997	5	31	14	4
47		March 13-April 2, 1997	13	22	13	0
48		April 3-25, 1997	6	39	10	21
49		April 26-May 14, 1997	15	28	12	5
50		May 15-June 13, 1997	15	35	22	8
51		June 14-30, 1997	5	25	15	6
52		July 1-18, 1997	6	19	10	2
53		July 19-Aug. 11, 1997	6	34	19	11
54		Aug. 12-29, 1997	6	25	18	5
55		Aug. 30-Sept. 19, 1997	4	37	18	17
56		Sept. 20-Oct. 12, 1997	7	36	24	5
57		Oct. 13-30, 1997	6	35	12	20
58		Nov. 1-24, 1997	2	9	7	0
59		Nov. 25-Dec. 4, 1997	18	50	41	4
60		Dec. 5-21, 1997	9	26	10	8

61		Dec. 22, 1997-Jan. 5, 1998	2	13	8	0
62		Jan. 6-26, 1998	4	22	9	2
63		Jan. 27-Feb. 13, 1998	5	22	13	0
64		Feb. 14-March 13, 1998	8	35	11	2
65		March 14-April 2, 1998	11	26	10	0
66		April 3-May 8, 1998	12	56	19	8
67		May 9-17, 1998	9	27	10	6
68		May 18-June 10, 1998	12	54	13	9
69		June 11-July 1, 1998	8	32	55	16
70		July 2-23, 1998	8	50	37	8
71		July 24-Aug. 12, 1998	9	68	28	33
72		Aug.13-Sept. 3, 1998	9	83	27	43
73		Sept. 4-23, 1998	11	36	20	12
74		Sept. 24-Oct. 2, 1998	6	22	22	0
75		Oct. 25-30, 1998	14	59	22	15
76		Oct. 26-Nov. 6, 1998	10	18	17	1
77		Nov. 7-26, 1998	13	82	25	54
78		Nov. 27-Dec. 15, 1998	12	129	30	92
Total			484	2154	1216	531

Note:

1. NMG = Nuclear Monitoring Group
2. The majority of sites were visited more than once during the 78 missions of IAEA's nuclear monitoring groups; therefore the total number of visits is sometimes greater than the total number of sites (both nuclear-related and unrelated) visited by each mission.

11. Termination and Abandonment of the Program

In order to address this subject, it would be necessary to classify the activities of the past Iraqi Nuclear Program (INP) after January 1991 into the following five categories (see 1.4 of the present CAFCD for more details):

I)   Research and Development (R&D).

II)   Design and various related engineering support.

III) Activities related to UNSCR 687 (1991).

IV)   Redefinition of missions.

V)   Documentation.

1. As for category (I), no R&D work related to the former INP (as described in this document) was carried out after January 17, 1991, by PC-3, EDC, DP or IAEC.

2. Activities carried out under category (II) included the following:

   a. Projects 602B and 603: After the destruction of the buildings that housed projects 601 and 602 at Tuwaitha, a verbal order was issued by HK, and then a written directive was issued by the director of PC-3 during Feb. 1991, to the design personnel concerned to redesign these projects at new site locations. This effort is adequately described in the present document. However, after Iraq Accepted UNSCR 687(1991), all design work ended.

   b. Spent fuel storage facility: The Tammuz-2 reactor building in Tuwaitha was severely damaged by aerial bombardment during Jan. 1991. The fuel elements of Tammuz-2, as well as the spent fuel elements from the July 14 (IRT-5000) reactor, were stored in the Tammuz-2 (underwater) storage facility. During Feb. 1991 all the fuel elements in the Tammuz-2 building were moved to location B and stored temporarily (underwater) in a number of pre-cast concrete water tanks. Eventually, all reactor fuel (fresh and spent) was removed from Iraq by the IAEA during the period 1991-1994.

During April/May 1991, a task was assigned to a design group to design and construct an appropriate alternative spent fuel wet storage facility. The site chosen was in Hamath at Tuwaitha. The basic design was completed, and civil construction started, and only the initial excavation was near completion when the project was terminated early in October 1991.

3. Category (III) includes all activities carried out by Iraq that were related to UNSCR 687 (1991). In particular, the movement and destruction of equipment and materials are adequately described in Addendum 1 Part 2 Annex 1 of this document. No further comments are necessary on this category except to point out that the destruction order for equipment and materials issued to the army early in July 1991 is a milestone indicating the abandonment of the former nuclear program.

4. The following activities are typical of what was carried out under the category (IV) after Feb. 1991:

   a. The engineering design and installation groups of PC-3 and Dijla and Al-Rabee factories were fully engaged after the war in the rehabilitation of the damaged infrastructural industrial establishments of Iraq, particularly the oil refineries and the electricity generation power stations.

   b. The senior chemical staff were assigned a task by the director of PC3 to propose new future missions for the various chemical groups of PC3 from March 1991 onwards. A directive was issued on August 18, 1991, to set-up a standing committee of leading research chemists and chemical engineers to work on the new missions defined in this directive (see section 1.4 of this document for more details).

   c. The mission of Al-Atheer center was redefined, as will be explained in 5 (b) and 5 (c) below. The new center set-up had the task of developing the production of materials.

   d. Chapter 2 (section 2.4) of the present declaration contains an organigram (Fig. 2.4.1) that gives complete details of all formerly attached entities to PC-3 and EDC. The full names of these entities, as well as their acronyms (as given in tables 2.4.1 & 2.4.2), are also listed. Moreover, a list of decrees, ministerial directives and orders totaling 66 are also listed. Their copies were handed over to the IAEA AT. These decrees, ministerial directives, and orders redefined the administrative links and missions of all the previously attached entities to the former PC-3 and EDC. These orders and decrees indicate that all the organizations associated with PC3 had been assigned new missions.

5. As for category (V), the proper documentation services that were available to PC3 before Jan. 17, 1991, were severely hampered by the destruction of facilities at Al-Tuwaitha. However, makeshift services were available from May/June 1991, and the following reports were issued:

   a. Basic design reports for 602B and 603 were issued in June 1991 and were discussed within category (II) above. These reports were handed over to the AT during IAEA-29.

   b. Report no. 99100100 prepared by Al-Atheer center on May 4, 1991, entitled "Report on the overall mission of Al-Atheer's departments for the development of the production of materials": This report redefined the new mission of Al-Atheer center as a "Materials R&D center" as a prelude to abandoning its former mission. It was submitted for approval by the PC-3 directorate, and after it was approved, a new version of the report described in (c) below was issued by the newly approved center (backdated to 1988) that included these new assignments.

   c. Report no. 856-01 issued by the "Materials R&D Center" on May 14, 1991, entitled "Feasibility Study for the Materials Center." It describes the new mission of Al-Atheer and the tasks assigned to its various departments after its former mission was abandoned.

   d. Report no. 99100200 prepared by the planning and follow-up department of "Al-Atheer Center for Developing the Production of Materials" issued on Sept. 10, 1991, entitled "Report on Achievements during the period June 1, 1990 - June 7, 1991".

The final conclusion on the subject of this note is now evident. No activities of any substance-related to the former INP were carried out during and beyond April 1991. All nuclear program activities were practically terminated and abandoned during April 1991, and only reports of previous accomplishments and new missions (non-proscribed) were issued later. Therefore, there is no substance to the allegation that INP activities were undertaken by PC-3, IAEC, EDC or their relevant support facilities during or after April 1991.

Furthermore, the points discussed below corroborate the fact that no activities of any substance relevant to the former mission of the INP were carried out beyond April 1991. They prove beyond any reasonable doubt that the former INP was abandoned altogether.

1. The decision of the government of Iraq to unilaterally destroy the equipment and materials early in July 1991 was a major milestone in the abandonment of the former INP.

2. The IAEA AT leader addressed a letter dated March 30, 1998, to Deputy Prime Minister Tariq Aziz requesting further clarification on this subject. The Deputy Prime Minister instructed the Under-Secretary of the Ministry of Foreign Affairs, Dr. Riyadh Al-Qaysi, to respond to this matter. In his letter of April 23, 1998 (attachment 1.5.1), Dr. al-Qaysi recalled to the attention of the IAEA AT that *"no governmental decree was issued at the time when the former INP was adopted and, therefore, the abandonment of the former INP required no such decree."* Also, he recalled the letter from the Minister of Foreign Affairs, Mr. M. S. Al-Sahaf dated May 1, 1997 (attachment 11.2) to the IAEA Director General reaffirming Iraq's obligations under NPT and the safeguards agreement signed with the IAEA. This reaffirmation fulfills Iraq's obligations stipulated in Para. 11 on UNSCR 687 (1991) and was also documented by the IAEA in GOV/2931 dated June 5, 1997, GC(41)/20 dated Sept. 9, 1997, and also in UNSC document S/1997/779 dated Oct. 8, 1997. Furthermore, the IT letters to the IAEA AT mentioned above were also recalled (August 26, August 31 and Sept. 3, 1997). These clarifications should be sufficient for proving that the past INP is well and truly terminated and abandoned.

3. In a meeting with Deputy Prime Minister Mr. Tariq Aziz on July 2, 1998, the IAEA AT leader again addressed the subject of abandonment of the former INP. Again, the Deputy prime Minister reiterated the points made in Dr. R. Al-Qaysi's letter,

April 23, 1998. These points were reiterated in a second letter dated July 16, 1998, from Dr. R. Al-Qaysi (attachment 1.5.3) (Under-Secretary of Foreign Affairs) to the IAEA AT leader. All these points should have been sufficient for the IAEA to make an unambiguous report to the UNSC stating that Iraq has complied fully with UNSCR 687 (1991) and 715 (1991) in the nuclear area.

Reference:

Letter from NMD to the IAEA AT dated 13/08/1997 entitled

"Abandonment of the program".

12. Concluding Remarks:

1. On April 6, 1991, Iraq accepted UNSCR 687 (1991).
2. On Nov. 26, 1993, Iraq accepted UNSCR 715 (1991). However, Iraq had practically implemented OMV activities since April 1992.
3. Iraq has provided all the necessary information to the IAEA AT relevant to discharging its mandate under the operative UNSC resolutions.
4. Iraq disclosed to the IAEA AT the status of all former INP related equipment and materials. It allowed the IAEA AT unrestricted access to all its facilities since 1991.
5. Iraq has cooperated fully with the IAEA and responded cooperatively and positively to all IAEA AT requests. It can safely be said that no outstanding matters of any substance-related to the past INP remain uncounted for today.
6. The Nuclear Material Balance is complete. All nuclear materials have been adequately accounted for (see para 72 of IAEA report S/1997/779).
7. The FFCD-F submitted by Iraq on Sept. 7, 1996, together with its addenda, has been considered to be adequate by the IAEA (see para 79 of IAEA report S/1997/779 dated Oct. 8, 1997).
8. All facilities, equipment, and materials of the former INP have been destroyed or rendered harmless. All dual-use equipment and

materials were subject to ongoing monitoring and verification (see para 72 of IAEA report S/1997/779).

9. All relevant former INP documentation has been handed over to the IAEA AT (see also para 57 of IAEA report S/1997/779).

10. The OMV plan has been up and running since 1994, and no proscribed equipment, materials or activities has been found or detected in Iraq, as confirmed by the results of a multitude of no-notice inspections carried out by IAEA inspectors from 1994 up to the present time (see para 82 of IAEA report S/1997/779). **The IAEA AT unilaterally suspended its OMV and all its other activities in Iraq on Oct. 29, 1997, and resumed these activities on Nov. 21, 1997, and then ceased all OMV activities after Dec. 15, 1998, when the resident NMG group departed from Iraq at the behest of the executive chairman of UNSCOM and without the approval of the UNSC.**

11. The IAEA AT stated that within the HHF cache, a file was found that contained correspondence between the Mukhabarat and Department 3000 of IAEC, which later became PC3. Included in this file was a few pages relating to an approach made by a foreign national, received in Oct. 1990, who offered assistance, for financial reward, in nuclear weapon design and in the procurement of any material that may be required.

Clearly, there is one original version of this file in existence today, and this version is in the sole possession of the IAEA AT. Therefore, the information contained in this file is authentic and up to date (until the abandonment of the former INP) and needs no further corroboration.

ON SEVERAL OCCASIONS, the IT pointed out to the IAEA AT that no external assistance was received by the former INP, other than that already declared to the IAEA AT and documented in the file in question.

This file which is in possession of the IAEA AT, is complete up to Jan. 1991, when the former INP was effectively abandoned, and no further transactions on this offer appear in the file. This corroborates the position of the IT that stated categorically that no assistance was received through this offer.

Furthermore, it may be prudent for the IAEA AT to seek verification of the IT's position from the foreign national and the foreign scientist named (both well known to the IAEA), as was suggested by the representative of a permanent member of the UNSC during the presentation of IAEA's report at the UNSC session on July 27, 1998.

12. Paragraph 16 of UN Security Council resolution 1051 (1996) requested the IAEA Director General to submit a consolidated report every six months to the Security Council, commencing on April 11, 1996.

13. The IAEA Director General submitted the fourth consolidated report on Oct. 8, 1997 that was circulated as document S/1997/779. This report was 95 pages long and in two parts.

13.1. Part One of the report (11 pages) describes the progress made by the IAEA during the six-month period April 1 - Oct. 1, 1997, in its implementation of the OMV plan and the discussions held with the IT on FFCD related matters pertaining to the former INP. The IT wishes to make the following comments on Part One:

    a.   In para 10 of this report, the IAEA claimed that *"The material and equipment found at the Tuwaitha Fire Station burial site is considerably less than Iraq has stated to have been buried at that location."* This claim by the IAEA is unfounded. During the General Conference of the IAEA held in Vienna during the period Sept. 22 - Oct. 3, 1997, the IAEA AT demanded from the IT to present the status of destroyed and buried equipment at the Tuwaitha site within 72 hours. Otherwise, it would report a negative position to the UNSC in its forthcoming October 1997 report! However, IT worked round the clock and mounted a large campaign in order to comply with IAEA AT's ultimatum. Many sites were excavated, and the buried equipment and materials that were destroyed more than seven years previously were uncovered and accounted for within the time span specified by the IAEA.

    b.   **The comment made by the IAEA AT in para 25 is incorrect.** The IT indicated precisely to the AT the state of development of the conceptual design of the device and its associated technologies.

c. **The comment made by the IAEA AT in para 26 is incorrect.** The IT provided detailed information, including ministerial orders and decrees defining new missions (including re-assignments) of former INP facilities. This IT position in this regard is presented in detail in section 11 of this summary.

d. The document referred to in para 27 was reviewed by the IT and presented to the IAEA AT in letter no. 7 of Aug. 13, 1997. **Therefore, the IT completed its task on this matter.**

e. The comments made by the IAEA in para 40 are incorrect. After Sept. 1995, the IT responded openly and in sufficient detail to any IAEA AT request (including those that were considered to be unreasonable). The IAEA AT's comments were made piecemeal; for example, FFCD related comments were received on five separate occasions over a period of nearly 17 months! They should have been presented once and for all a few weeks after the first FFCD draft was submitted on March 1, 1996.

f. In para 43, the IAEA indicated that it made repeated requests to the IT to locate device-related destroyed equipment. In fact, upon the continued insistence of the IT, the IAEA AT specified in May 1997 that certain device-related items needed to be verified. The IT made concentrated efforts at locating these items, and an adequate status was presented to the IAEA AT in July 1997 that closed this matter.

13.2. Part Two of the report comprised 84 pages and **is considered to be the final report of the IAEA on the disarmament phase.** It includes the following:

a. "overview of IAEA activities regarding identification and destruction, removal and rendering harmless of Iraq's capabilities related to nuclear weapons" (11 pages).

b. Attachment 1 "The components of Iraq's clandestine nuclear program" (40 pages).

c. Attachment 2 "Chronology of major events" (7 pages).

d. Attachment 3 "Destruction, removal and rendering harmless" (5 pages).
e. Attachment 4 "Summary of IAEA inspection campaigns" (21 pages).

14. The IT wishes to make the following comments on the first 11 pages of Part Two:
a. Para 67 implies that the amount of additional documentation requested by the IAEA are similar to the amount the IT volunteers. In fact, the amount of documentation volunteered by the IT by far exceeded (probably by orders of magnitude) the amount requested by the IAEA.
b. In para 74, the IAEA claims that even after August 1995, the IT *"continues to limit the scope of information provided in response to IAEA questioning in an effort to understate the capability etc.".* **This statement is false and is not corroborated by any factual or specific example.**
c. In para 77, the IAEA stated that *"However, no documentation or other evidence is available to show the actual status of the weapon design when the programme was interrupted."* The IT wishes to point out that this R&D program was practically terminated on Jan. 17, 1991, and was never resumed thereafter.

15. In spite of the tremendous efforts made by the IT during the period 1991- 1998, the IAEA failed to report to the UNSC that Iraq has completed all actions contemplated in UNSCR 687 (1991) in the Nuclear Area. On a number of occasions during 1998, the IAEA AT leader promised to state this finding in IAEA's report to the UNSC but failed to live up to these promises. Moreover, the IAEA reports always contained a few minor *"unfinished questions"* that are readily picked up by those that have no intention to bring this matter to a close in the UNSC.
16. It is evident from the presentation made in this Extended Summary that "Iraq has completed all actions contemplated in UNSCR 687 (1991) in the Nuclear Area".

Table 8.2
Detailed Clarifications Submitted by the Iraqi Team to the IAEA - Action Team
Following the IAEA Action Team Visit to Baghdad During July 19-23, 1997 and Action Team Letters of ATL1 - ATL7

Letter No.	Action Team Request (Heading)	Iraqi Counterpart Response Date of issue	No. of pages	No. of attachments
1.	EDC equipment in Amman (point #13 of CM+)	04/08/1997	1	2
2.	Procurement and external assistance / ATL1 and (point #9 of CM)	09/08/1997	2	0
3.	Cooperation with IAEA (point #15 of CM)	09/08/1997	1	0
4.	Comments on Addendum II part 1 of the FFCD-F of Sept. 7, 1996	11/08/1997	1	0
5.	Nature of the HHF cache (point #1 of CM)	11/08/1997	4	0
6.	Nature of the HHF cache (point #1 of CM) further note	13/08/1997	1	1
7.	Movement, concealment and destruction of materials and documents (point #3 of CM)	13/08/1997	3	1
8.	Translation of the list of Al-Kawther technical reports (point #4 of CM)	13/08/1997	1	1
9.	Status of the streak camera system (point #8 of CM)	13/08/1997	1	1
10.	Information on report that deals with uranium purification (point #12 of CM)	13/08/1997	1	0
11.	Locations of Al-Atheer equipment (point #16 of CM)	13/08/1997	2	2
12.	Movement and destruction of documentation (point #7 of CM)	14/08/1997	1	1
13.	Procurement / ATL1	19/08/1997	1	0
14.	Request for Action Team leader visit to Baghdad	19/08/1997	1	0
15.	Sites 4 & 5 of EDC (point #6 of CM)	20/08/1997	1	Some debris found at location no. 4
16.	List of directives and decrees indicating the abandonment of the former INP and the organigram of PC-3 and EDC (point #5 of CM)	26/08/1997	1	4
17.	Status of the nuclear weapon / ATL1 and point #2 of CM	27/08/1997	7	5
18.	Abandonment of the program / ATL1	31/08/1997	4	2
19.	Actions and reactions related to concealment and unilateral destruction of materials, equipment and documentation / ATL1	03/09/1997	6	0
20.	Belarus contract / ATL2	04/09/1997	1	0
21.	Status of some G4 equipment / ATL3 and point #10 of CM	11/09/1997	1	1

Table 8.2 (continued -1)

Letter No.	Action Team Request (Heading)	Iraqi Counterpart Response		
		Date of issue	No. of pages	No. of attachments
22.	The addition of a new chapter to FFCD-F	15/09/1997	3	0
23.	External assistance to the former Iraqi Nuclear Program (INP)	15/09/1997	3	1
24.	Detailed clarifications to all AT remarks	16/09/1997	1	1
25.	The addition of new chapter to FFCD-F	22/09/1997	2	0
26.	Response to the AT letter of Sept. 19, 1997	22/09/1997	2	0
27.	Procurement/MIC-Belarus	26/09/1997	1	1
28.	AT letter of Sept. 27, 1997 /ATL5	01/10/1997	2	1
29	Action Team letter of Oct. 1, 1997/ATL6	06/10/1997	1	0
30	Cross reference of Excavated Equipment	07/10/1997	1	1
31	Head of Iraqi Delegation to IAEA GC 41	08/10/1997	3	0
32	AT letter October 4, 1997	13/10/1997	1	3
33	Procurement / MIC	13/10/1997	1	2
34	Tritium issues NMG-9714/NMD/Req-001	14/10/1997	1	0
35	Diffusion pumps NMG-9714/NMD/Req-002	15/10/1997	1	1
36	Follow-up action - IAEA 30-5 related to G4 equipment	15/10/1997	1	1
37	Planned visit of Mr. G. Dillon to Iraq	31/10/1997	1	0
38	Monitoring Activities	03/11/1997	1	0
		Total	**68**	**33**

+ CM   = List of Remarks Presented During the Closing Meeting of July 23, 1997.

++ATL1 = Action Team Letter of August 1, 1997.   ATL4 = Action Team Letter of Sept. 19, 1997.
ATL2 = Action Team Letter of August 26, 1997.   ATL5 = Action Team Letter of Sept. 27, 1997.   ATL7 = Action Team Letter of Oct. 4, 1997
ATL3 = Action Team Letter of Sept. 6, 1997.   ATL6 = Action Team Letter of Oct. 1, 1997